Disposal and Management of Solid Waste

Pathogens and Diseases

Disposal and Management of Solid Waste
Pathogens and Diseases

Eliot Epstein

CRC Press is an imprint of the
Taylor & Francis Group, an **informa** business

CRC Press
Taylor & Francis Group
6000 Broken Sound Parkway NW, Suite 300
Boca Raton, FL 33487-2742

© 2015 by Taylor & Francis Group, LLC
CRC Press is an imprint of Taylor & Francis Group, an Informa business

No claim to original U.S. Government works

Printed on acid-free paper
Version Date: 20140922

International Standard Book Number-13: 978-1-4822-4814-2 (Hardback)

This book contains information obtained from authentic and highly regarded sources. Reasonable efforts have been made to publish reliable data and information, but the author and publisher cannot assume responsibility for the validity of all materials or the consequences of their use. The authors and publishers have attempted to trace the copyright holders of all material reproduced in this publication and apologize to copyright holders if permission to publish in this form has not been obtained. If any copyright material has not been acknowledged please write and let us know so we may rectify in any future reprint.

Except as permitted under U.S. Copyright Law, no part of this book may be reprinted, reproduced, transmitted, or utilized in any form by any electronic, mechanical, or other means, now known or hereafter invented, including photocopying, microfilming, and recording, or in any information storage or retrieval system, without written permission from the publishers.

For permission to photocopy or use material electronically from this work, please access www.copyright.com (http://www.copyright.com/) or contact the Copyright Clearance Center, Inc. (CCC), 222 Rosewood Drive, Danvers, MA 01923, 978-750-8400. CCC is a not-for-profit organization that provides licenses and registration for a variety of users. For organizations that have been granted a photocopy license by the CCC, a separate system of payment has been arranged.

Trademark Notice: Product or corporate names may be trademarks or registered trademarks, and are used only for identification and explanation without intent to infringe.

Library of Congress Cataloging-in-Publication Data

Epstein, Eliot, 1929-
  Disposal and management of solid waste : pathogens and diseases / author, Eliot Epstein.
     pages cm
  Summary: "In developed countries wastewater and sewage sludge are disposed by means that reduce or minimize exposure by humans to disease organisms. Most municipal solid waste goes to landfills which have liners to protect ground water. Humans are often exposed to pathogens, resulting in serious diseases from the disposal of human and animal wastes. This book describes the various pathogens and diseases that can be found in solid waste and describes the means and opportunities for disposal and management of various solid waste materials"-- Provided by publisher.
  Includes bibliographical references and index.
  ISBN 978-1-4822-4814-2 (hardback)
  1. Sanitary microbiology. 2. Pathogenic microorganisms. 3. Refuse and refuse disposal--Heatlth aspects. 4. Refuse and refuse disposal--Toxicology. 5. Sewage--Microbiology. 6. Diseases--Causes and theories of causation. I. Title.

QR48.E67 2015
579'.165--dc23   2014033131

Visit the Taylor & Francis Web site at
http://www.taylorandfrancis.com

and the CRC Press Web site at
http://www.crcpress.com

Dedication

I am dedicating this book to the people in developing countries who need to be better informed on diseases and their prevention and to use whatever means in their possession to eliminate illnesses and provide better public health.

Contents

Preface xi
Acknowledgments xv

Chapter 1   Pathogens and Diseases Associated with Disposal and
Management of Solid Waste 1

    Introduction 1
    Definitions 5
    Sources of Pathogens in Solid Waste 6
    Infection of Humans from Solid Waste 8
    Summary 10
    References 11

Chapter 2   Risk and Exposure 13

    Introduction 13
    Risk Assessment 16
    Hazard Assessment and Identification 17
    Dose–Response Assessment 18
    Exposure Assessment 18
    Risk Characterization 19
    References 19
    Suggested Reading 20

Chapter 3   Odors as a Health Issue 21

    Introduction 21
    Basic Concepts 21
    Odor and Health 26
    Summary 28
    References 28

Chapter 4   Pathogens and Diseases of Solid Wastes: Municipal Solid Waste
or Garbage 29

    Introduction 29
    Pathogens and Their Diseases 33
    Vectors 37
    Summary 38
    References 39

Chapter 5	Pathogens and Diseases of Sewage Sludge, Septage, and Human Fecal Matter 41	
	Introduction 41	
	Pathogens and Diseases... 44	
	References 46	
Chapter 6	Pathogens in Soils 47	
	Introduction 47	
	Bacteria.. 54	
	Viruses... 57	
	Parasites... 58	
	Summary 59	
	References 59	
Chapter 7	Geophagy and Human Pathogens in Plants 63	
	Introduction 63	
	Geophagy 63	
	Pathogens on Plants... 67	
	Summary 70	
	References 70	
Chapter 8	Bioaerosols 73	
	Introduction 73	
	Fungi and Pathogens Commonly Found in Outdoor and Indoor Environments 76	
	Aspergillus fumigatus 79	
	Endotoxin 80	
	Glucans 81	
	Actinomycetes 81	
	Mycotoxins 82	
	References 83	
Chapter 9	Pathogens in Animal Waste and Manures 87	
	Introduction 87	
	Pathogens and Diseases Transmitted to Humans.................... 89	
	Bacteria... 90	
	Viruses... 90	
	Parasites (Protozoans) 90	
	Summary 93	
	References 94	
Chapter 10	Pathogens in Food and Water... 97	
	Introduction 97	
	Foodborne Diseases 98	

Contents

    Waterborne Diseases 105
    Summary 109
    References 109

Chapter 11 Disposal and Management of Solid Waste 111
    Introduction 111
    Landfills or Dumps 112
    Incineration and Burning 112
    Anaerobic Digestion 113
    Land Application 113
    Composting 114
    Lime Stabilization 115
    Summary 115
    References 116

Appendix: Details of the Pathogens and Their Diseases 117
    Actinobacillus pleuropneumoniae 117
    Acinetobacter baumannii 117
    Actinomyces spp. 117
    Aeromonas spp. 118
    Ascaris lumbricoides 118
    Bacillaceae spp. 119
    Bacillus spp. 120
    Bordetella spp. 122
    Campylobacter jejuni 122
    Cellulomonas 124
    Entamoeba histolytica 124
    Enterobacteria 126
    Escherichia coli 126
    Klebsiella pneumonia 127
    Micrococcus 128
    Mycobacterium spp. 129
    Neisseria 131
      Neisseria meningitidis 131
      Neisseria gonorrhoeae 132
    Proteus spp. 133
    Pseudomonas spp. 133
    Salmonella spp. 135
    Serratia plymuthica 137
    Staphylococcus aureus 138
    Streptococcus faecalis (*Enterococcus faecalis*), *Streptococcus
    pyogenes*, and *Streptococcus pneumoniae* 139

Index 143

Preface

Pathogens in solid waste and many of their diseases result from the disposal and management of those same solid wastes. In developed countries, sanitation and management of solid waste are highly controlled or regulated by local or state entities. The disposal of solid waste in developing countries is predominantly through landfills or incineration. Environmentally safe systems such as composting are relatively rare. In the United States, the exception involves human wastes after collection and treatment that result in wastewater sludge. Land application and composting of wastewater sludge is the preferred disposal management system in the United States.

This is not the case in many developing countries. Collection and disposal of solid waste are poor and neglected. Human and animal wastes are often left on the surface adjacent to residences. George (2008) points out that in India nearly 800 million persons spread contagious diseases in this fashion. Even where latrines were installed, they were not used. In developing countries, dumps and outdoor defecation can be a major source of pathogens and bioaerosols. Not only workers are at risk but also the general population. George (2008) indicates that in India there are between 400,000 and 1.2 million manual scavengers, who are not only affected directly by pathogens and bioaerosols but also by pathogens and dust containing bioaerosols brought into the home.

In developed countries, pathogens and the diseases resulting from them are predominantly waterborne or foodborne. There have been relatively few illnesses reported in the literature from the disposal and management of sewage sludge, biosolids, or municipal solid waste (MSW). The populations most likely to be exposed are workers. Even diseases from direct contact or exposure during wastewater treatment, collection of MSW, or disposal of MSW are rare. Hygiene, sanitary conditions, and the education of workers are high. However, the potential exposure to diseases by workers can result from inhaling bioaerosols and having respiratory complications. Bioaerosols produced during composting of sewage sludge or MSW has been a concern for workers and populations residing near composting facilities. The bioaerosols of predominant concern have been *Aspergillus fumigatus* and endotoxins. Numerous studies have been conducted on this issue in the United States and United Kingdom.

In the United States, composting of sewage sludge and yard waste has increased significantly. Today, the disposal or utilization of sewage sludge is primarily through land application, followed by composting and landfilling. Bans from disposing of yard material or green waste in landfills resulted in a massive growth of outdoor windrow composting. This resulted in concern for exposure to bioaerosols for workers and residents living near composting facilities. I have been involved in lawsuits by residents against municipalities because the residents feared exposure to *Aspergillus fumigatus*, a fungus predominating during composting since it survives the high temperatures during composting.

Bioaerosols in the indoor environment have been a source of respiratory illnesses such as asthma. Molds and mildew are found in air conditioning units, humidifiers, laundry rooms, plumbing leaks, and other sources of moisture. The "sick building" syndrome has been attributed to bioaerosols.

This book devotes two chapters to bioaerosols. One chapter covers the general aspects, including the indoor environment, bioaerosols from MSW facilities, as well as measuring and enumerating technologies. The second chapter is devoted to composting. This is not only because of the vast literature but also because composting can represent a viable low-cost technology to deal with the vast production of solid waste and its disposal.

In 1931, Sir Albert Howard and Yeshwant Wad published *The Waste Products of Agriculture*, addressing a composting method called the Indore method, a technique to produce stable organic matter termed humus as a means of promoting crop growth. Although this was applied to agricultural waste, it also had an application to solid waste. This aspect was never endorsed. A later book, *An Agricultural Testament*, indicated that the Indore method had become widespread. The organic matter and fertility were the benefits. I describe this and other newer, low-cost technologies. In developing countries, composting by any means can be a viable technique for disease prevention and control (Howard 1935).

Anaerobic digestion of human wastes as well as other wastes can be greatly expanded, especially in developing countries. Anaerobic digestion provides methane, which can be a source of energy. The solids remaining can be used either directly as a source of nutrients and organic matter for plant growth or composted for further disinfection and use in horticulture, public works, and agriculture. In India, I learned of small anaerobic digesters used in the home to provide methane gas for cooking. In the United States, anaerobic digestion of biosolids takes place in large expensive digesters. Methane from these facilities is used for power generation and even power for vehicles. These expensive digesters are not a method to handle sewage sludge in developing countries since paying off the capital investment and operations through taxation would be prohibitive. However, I saw several inexpensive anaerobic digesters used for agricultural and food waste that could be adapted for other wastes in developing countries. According to George (2008), 15.4 million rural households in China have biogas (methane from anaerobic digestion) for cooking and other uses. She stated that Nepal has more digesters per capita than China. Covered lagoons have also been used to produce biogas from dilute waste from animal production facilities (Bowman 2009). In Minnesota, at a food-processing facility, I also saw a covered lagoon that produced methane.

This book describes the various pathogens that are found in solid wastes and the diseases they produce, with emphasis on their relation to developing countries. Sanitation is highly neglected in developing countries. More money is often spent on arms and tribal warfare than on sanitation and food production. In Iran and Egypt, the military controls many industries and commercial enterprises.

Besides better sanitation, hygiene and education that promotes disease prevention are extremely important. This enhances prevention of waterborne and foodborne diseases.

REFERENCES

Bowman DD. 2009. *Manure Pathogens*. Alexandria, VA: WEF Press.
George R. 2008. *The Big Necessity*. New York: Metropolitan Books.
Howard A. 1935. The manufacture of humus by the Indore process. *J Royal Soc Arts* 74: 26–60.
Howard A. 1943. *An Agricultural Testament*. New York: Oxford University Press.
Howard A and Wad YD. 1931. *The Waste Products of Agriculture: Their Utilization as Humus*. London: Oxford University Press.

Acknowledgments

I would like to thank my family: Esther, my wife; and my three children, Beth, Jonathan, and Lisa, for their encouragements, suggestions, and patience. My wife worked at the National Institutes of Health. All three of my children are involved in human health and welfare. Beth is an occupational therapist, often working with autistic children. Jonathan is director of surgical pathology at Johns Hopkins Hospital and is a professor of pathology, urology, and oncology involved in work with prostate cancer and other urological problems. Lisa works as a nutritionist for an organization called WIC (Women, Infants, and Children). Their dedication to human health encouraged me to pursue this direction.

1 Pathogens and Diseases Associated with Disposal and Management of Solid Waste

INTRODUCTION

This book covers:

- Global aspects of pathogens and diseases from solid waste
- Pathogens and diseases in various solid wastes other than hospital wastes
- Disposal and management of solid wastes in relation to diseases

Most of the scientific data available relate to foodborne and waterborne pathogens and their diseases. This primarily involves ingestions or direct intake into the gastrointestinal system. With solid waste, dermal and respiratory infections and diseases are important. Probably, as will be shown, respiratory diseases are the most important route for diseases to workers and the public from solid waste. Pathogenic bioaerosols are also an important source of disease.

In developed countries, sanitation is often well developed. There is great emphasis on hygiene and general cleanliness. Water is treated with disinfectants such as chlorine, or water treatment facilities filter the water. Many homes also use water filters for the entire home, under the sink, or in closed containers.

This is not the case in developing countries. Although emphasis is placed on clean water, there is neglect of sanitation and management and disposal of solid waste. This is particularly true in small villages. In 1945, when I was in Cairo, Egypt, garbage of municipal solid waste (MSW) was piled in the streets. During the 1970s, I was invited to Milan, Italy, to suggest opportunities for management of MSW. During my stay, there was a strike of municipal garbage collectors. Garbage gathered in the streets. I am sure if water was not available or if homes could not dispose of sewage, a great cry would have taken place, and the problems would have been immediately solved, but disposal of MSW was of secondary importance. Few people were concerned with vectors and their impact on human health.

In recent years, there have been numerous instances of water and food contamination resulting from fouling by wild or domestic animals. In developing countries, especially in rural areas of Africa, India, and China, human waste disposal is a major concern, especially as related to health. Developing countries today account for 85.4% of the world's population, and an estimated 500,000 people defecate in the open. Poor sewage conditions often lead to a host of diseases and mortality. It is reported that poor sanitation kills 12 million people each year due to mosquito-borne diseases and through diseases from water contamination. The World Health Organization (WHO) has reported that 3.4 million people die as a result of water-related diseases that are often the result of poor sanitation and water contamination from raw sewage and human fecal matter. In villages and small communities, there is a direct relation between water contamination and sewage or solid waste disposal. Water resources such as wells and streams are downstream from where defecation or disposal of contaminated solid waste occurs. Sanitation is nonexistent. As will be shown, the role of vectors, especially mosquitoes, is extremely important in diseases and human health. The effort for eradication has been concentrated on medication and screening. Little effort is given to control the source of mosquitoes and their breeding. This is often directly related to solid waste management. Water can accumulate in numerous MSW sources such as tires, containers, discarded shoes, and so on. Proper sanitation includes restricting runoff that carries fecal matter, providing adequate drainage and water removal, providing adequate conditions for disposal of human waste, and least but not less important is providing education. In *The Big Necessity*, Rose George stated, "Sanitation is one of the best investments a country can make." Further, she asserted, "Excreta disposal is a political football, kicked between departments" (George 2008).

In developing countries where sewer systems and treatment are unavailable, often outdoor defecation is the predominant mode of fecal matter disposal. George (2008) points out that in India nearly 800 million persons spread contagious diseases in this fashion. Even where latrines were installed, they were not used.

From more than 3,500 years ago, the biblical portion of Deuteronomy 23 states, "You shall have a shovel in addition to your weapons, and it will be that when you sit outside, you dig with it; you shall go back and cover your excrement." Proper hygiene required the Israelites to defecate between 1,000 and 2,000 cubits or 1,000 to 4,000 feet away from the camp (George 2008). It was evident that they realized then that open defecation could result in diseases.

Foodborne diseases are a major problem. Contaminated food causes approximately 1,000 reported disease outbreaks, with an estimated 48 million illnesses, 128,000 hospitalizations, and 3,000 deaths in the United States annually (Gilliss et al. 2011). Some of these occurrences were related to solid waste, such as contamination by animal manure.

Water contamination can be related to sanitation. Two early examples of water contamination incidences are: (1) A Milwaukee, Wisconsin, outbreak with *Cryptosporidium* believed to be from animal or human waste contamination. This resulted in 100 deaths and 400,000 persons becoming ill (MacKenzie et al. 1994). (2) Another incident occurred in Walkerton, Canada. Starting May 15, 2000, many residents of the town of about 5,000 began to simultaneously experience purple diarrhea, gastrointestinal infections, and other symptoms of *Escherichia coli* infection.

At least 7 people died directly from drinking the water contaminated with *E. coli*, and about 2,500 became ill (Krewski et al. 2002).

The major foodborne illnesses are caused by the following pathogens:

- *Clostridium botulism*
- *Campylobacter*
- *Cryptosporidium*
- *Giardia*
- *E. coli* O157:H7
- *Listeria*
- Molds
- Noroviruses
- *Yersinia*
- *Salmonella*
- *Shigella*
- *Toxoplasma gondii*
- *Vibrio parahaemolyticus*

WHO estimated that in 2005 alone, 1.8 million died from diarrheal diseases. Many of these diseases were related to food and drinking water (WHO 2007). WHO conservatively estimated that the global death toll from diarrheal diseases is from 1.7 to 2.5 million deaths per year. Most of these occur in children under five years of age (WHO 2009). In many developing countries with poor resources, other diseases, such as cholera, shigellosis, and typhoid, take a large toll. On February 23, 2013, the *New York Times* reported that in Haiti over 2 years ending in 2012, 8,000 persons died and 646,000 were ill from cholera as a result of poor sanitary conditions.

The principal agents of diarrheal diseases are

- *Vibrio cholerae*
- *Salmonella* spp.
- *Shigella* spp.
- *Campylobacter* spp.
- *Escherichia coli* strains
- *Staphylococcus aureus*
- *Clostridium perfringens*
- *Clostridium difficile*
- *Giardia lamblia*
- *Cyclospora*
- *Cryptosporidium parvum*

A number of viruses, including enteric adenovirus, astrovirus, and calicivirus, can cause diarrhea.

Furthermore, fungal infections and airborne aspects related to fungi such as *Aspergillus*, molds, and actinomycetes have increased. In the United States, asthma affects over 25 million persons, 7 million of them children. Worldwide, it is estimated

TABLE 1.1
Deaths from Several Diseases in Developing Countries

Causes of Death	1990[a]	2002[b]	2008[a]
Lower respiratory infections	3,894,000	2,806,000	1,050,000
Diarrheal diseases	2,865,000	1,535,000	760,000
Malaria	928,000	1,246,000	480,000
Tuberculosis	1,978,000	961,000	400,000

[a] World Health Organization, Media Centre. 2014. The top 10 causes of death. May. Fact sheet 310. http://www.who.int/mediacentre/factsheets/fs310/en/index.html

[b] Parliamentary Office of Science and Technology. 2005. Post note, number 241. June.

that over 300 million people are affected. Indoor air pollution from molds or noxious fumes can result in allergies and asthma ("Asthma Statistics" 2011). The lack of hygiene, inaccessibility to soap and water, and lack of public health education are a major source of the problem. It is estimated that correction of these aspects could reduce diarrheal diseases by 50% (Centers for Disease Control and Prevention [CDC] 2010).

Table 1.1 shows deaths in developing countries from several diseases. The data were hard to obtain because of inconsistency in reporting. Lower respiratory tract infections, diarrheal diseases, and tuberculosis decreased for the three periods reported. Malaria increased from the reported years of 1990 to 2002 and then decreased in the reported year of 2008. Diarrheal diseases and malaria can result of poor sanitation and solid waste disposal as indicated in Chapter 4.

Infectious diseases are the greatest killers of adults and children. They account for 13 million deaths a year. Six diseases cause 90% of infectious disease deaths (Brundtland 1999). With children under five years old, acute respiratory illnesses are the leading cause of death. Lower respiratory tract infections refer to pneumonia, influenza, bronchitis, and bronchiolitis caused by common microbes with worldwide distribution. Malaria is the most serious parasitic disease in the world. Eradications are infeasible. Parasitic diseases are also a major cause of disease and death. Protozoal parasites of African trypanosomiasis cause African sleeping sickness, and American trypanosomiasis causes Chagas disease, with annual mortalities estimated at 55,000 and 25,000, respectively. Another protozoan disease, leishmaniasis, is highly fatal. Other causes as reported by WHO and the Centers for Disease Control and Prevention (CDC) are: amoeba, 54,000 deaths; hookworm, 90,000 deaths; *Ascaris*, 60,000 deaths; schistosomiasis, 38,000 deaths; and onchocerciasis, 30,000 deaths ("How the Other Half Dies" n.d.). Although these numbers will not be the same currently or each year, they still represent the situation in the sub-Saharan region of Africa and other developing countries.

DEFINITIONS

It is important to define the parameters discussed in this book. The three principal parameters are

- Pathogens
- Diseases
- Solid waste

Stedman's Medical Dictionary (1976) defines a pathogen as any virus, microorganism, or other substance causing disease. It further defines an opportunistic pathogen as one that is capable of causing a disease when the host defense mechanism (immunosuppression) is lowered. A disease is defined as an illness, sickness, interruption, cessation, or disorder of body functions, systems, or organs.

Solid waste includes the following:

- MSW or garbage
- Sewage sludge and biosolids (treated sewage sludge)
- Septage
- Manure
- Human excrement from domestic sources that do not have municipal disposal facilities

Solid waste can consist of MSW (garbage), which may contain discards of pet feces, contaminated tissues from humans, contaminated food waste (e.g., chicken with *Salmonella*), and other contaminated sources. In the United States, most of the MSW is landfilled. Incineration or heat recovery systems are few. There are approximately 200, mostly small, biosolid composting facilities in the United States. The largest facility is the Inland Empire composting facility located in Chino, California. There are few MSW composting facilities. Food waste collection and composting are increasing. Approximately 154 municipalities are collecting food waste. On June 17, 2013, Mayor Bloomberg of New York indicated that the city will initiate food waste composting as being done in several other cities.

In the United States, sewage sludge, the residue after anaerobic digestion, is frequently applied to the land. For sewage sludge to be applied to land, it must undergo treatment to stabilize and destroy some of the pathogen concentration. Sewage sludge that is treated is then classified as biosolids. Lime treatment is also a method of stabilizing sewage sludge prior to land application. In the United States, approximately 50% of biosolids are applied to land.

Composting of sewage sludge has grown significantly in the United States, with one of the largest facilities handling several hundred tons per day. The state of California has been the most progressive in utilizing composting of sewage sludge or biosolids.

Since excellent treatment of pathogens in manure has occurred recently (Bowman 2009), Chapter 9 is devoted to this subject to update the information.

SOURCES OF PATHOGENS IN SOLID WASTE

The sources of pathogens in solid waste are

- Discarded raw food
- Diapers
- Animal and pet wastes
- Discarded medicinal items such as Band-Aids, contaminated gauze, and the like
- Contaminated paper towels, nose wipes, and so on
- Wastewater containing excreta and urine
- Other domestic sources not using sewage systems (e.g., outhouses, pits, soil surfaces)
- Health care wastes
- Manure on land and runoff

Solid waste can contain pathogens such as viruses, bacteria, parasites, fungi, and actinomycetes. As an example, *Aspergillus fumigatus* can infect humans directly through the respiratory system. Airborne pathogens can land on crops and eventually infect foods and humans. Ritter et al. (2002) stated that, on a global scale, pathogenic contamination of drinking water poses the most significant health risk to humans. Contaminated water resulting from leachate from landfills or other disposal methods can enter drinking water supplies in ground or surface waters. Sobsey (1978) found enteric viruses in landfill leachate. Contaminated water can directly affect the food supply. Pathogens can enter the soil from applications of manure and sewage sludge (unless disinfected by composting, heat treatments, or lime). The pathogens can move through the soil (e.g., via fissures, wormholes, etc.) and eventually end in groundwater and drinking supplies (Ritter et al. 2002). Teschke et al. (2010) indicated that studies of endemic diseases across multiple water or sewage systems are rare.

The transmission route of pathogens from solid waste to the environment is shown in Figure 1.1. It is evident that human exposure to pathogens from solid waste is ubiquitous. A major source of diseases to children in developing countries is ingestion of soil. This is discussed under the topic of geophagia.

The soil and its environment are depicted in Figure 1.2. Chapter 6 is entirely devoted to pathogens in soil. The soil is a complex medium consisting of chemical, physical, and biological parameters. Soil water affects plant growth and the biological properties. It affects the survival of plants, as well as the microbial population. Soil water dissolves chemicals, as well as transports them through the soil. Plant nutrients assist in the growth of plants. The physical and chemical properties are inert, whereas the biological properties are dynamic.

It is well known that the soil may harbor pathogenic microorganisms as well as other hazardous material (Berg, Eberl, and Hartman 2005; Baumgardner 2012). Many of the pathogens may result from sewage sludge or biosolid application in developed countries, whereas in developing countries solid waste, raw sewage, or fecal matter disposal can contaminate soil. Helminths or geohelminths, such as

Pathogens and Diseases Associated with Disposal of Solid Waste

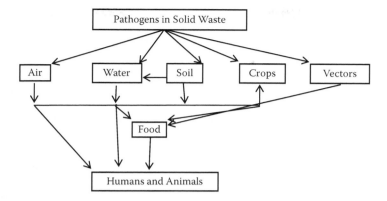

FIGURE 1.1 The transmission route of pathogens from solid waste to humans.

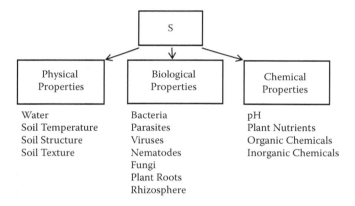

FIGURE 1.2 The physical, biological, and chemical properties of soil.

roundworm (*Ascaris lumbricoides*), whipworm (*Trichiura trichiura*), hookworms (*Ancylostoma duodenale* and *Necator americanus*), and threadworm (*Strongyloides stercoralis*), can result in infection from soils (Brooker, Clements, and Bundy 2006). WHO estimates that as many as 1 billion persons are currently infected by soil-transmitted helminths (WHO 2011).

Tetanus and botulism are two pathogenic organisms found in soil. The causative organisms are *Clostridium tetani* and *Clostridium botulinum*, respectively. *C. tetani* is distributed worldwide and can be found in fields, playgrounds, schoolyards, and other locations (Baumgardner 2012; Ebisawa et al. 1986). It produces a life-threatening toxin affecting the muscle.

C. botulinum is also found in soil and water (Baumgardner 2012). Its toxin produces paralysis and on occasion fatal illness. The April 28, 2013, *New York Times* magazine reported an unusual case of an infant apparently contaminated from soil by the organism *C. botulinum*. If not for proper diagnosis, the child probably would have died.

Soil can either directly or indirectly be detrimental to human health. This can occur in ingestion, inhalation, and dermal absorption (Abrahams 2002). Direct ingestion of soil particles, geophagia (discussed further in Chapter 7), also termed pica (e.g., pica child), meaning one who eats soil, can take place in both developed and developing countries (George and Ndip 2011; Vermeer and Frate 1979). I am aware of one incident that occurred when a young female child played in a sandbox where the previous year a young boy relative played with her in the same sandbox. The boy was from a Caribbean country, and while playing in the box, he defecated. The girl became very ill, and after visiting several doctors and hospitals, she died. Apparently while playing the following year, she ingested some of the contaminated sand. A parasite was able to get into her bloodstream and eventually into the brain. Although this incidence is rare, geophagia occurs in both developed and developing countries, with the majority of the occurrences in developing countries (Vermeer and Frate 1979). George and Ndip (2011) reported that in rural South Africa there was a widespread practice of geophagia among young girls. This subject is discussed in detail in Chapter 7.

The potential for inhalation from the soil surface is low. However, the application of wastes and surface dumping of trash can result in inhalation of bioaerosols that enter the respiratory system (see Chapter 8). Aerosolization of dust can contain numerous organisms, including bacteria and fungal spores (Sing and Sing 2010). Coccidioidomycosis, commonly known as "valley fever," as well as "California fever," "desert rheumatism," and "San Joaquin Valley fever," is a fungal disease caused by *Coccidioides immitis* or *Coccidioides posadasii* and is endemic in certain parts of Arizona, California, Nevada, New Mexico, Texas, Utah, and northwestern Mexico (Rapini et al. 2007).

C. immitis, a fungus, resides in the soil in certain parts of the southwestern United States, northern Mexico, and parts of Central and South America. It is dormant during long dry spells, and then develops as a mold with long filaments that break off into airborne spores when the rains come. The spores, known as arthroconidia, are swept into the air by disruption of the soil, such as during construction, farming, or an earthquake. Infection is caused by inhalation of the particles. The disease is not transmitted from person to person. The infection ordinarily resolves, leaving the patient with a specific immunity to reinfection. *C. immitis* is a dimorphic saprophytic organism that grows as a mycelium in the soil and produces a spherule form in the host organism.

Polymenakou et al. (2008) reported on the short distance of areal transport of a dust-borne pathogen, *Neisseria meningitidis*, which caused seasonal outbreaks of bacterial infection from that organism in Africa.

INFECTION OF HUMANS FROM SOLID WASTE

There are several potential sources of human infections from solid waste. As indicated, solid waste applied or deposited on soil can result in pathogen transmission to humans.

Pathogens and Diseases Associated with Disposal of Solid Waste

FIGURE 1.3 Individuals sorting usable material from municipal solid waste in Wuhan, China.

The potential sources are as follows:

- Direct contact with pathogens present in solid waste
- Aerosolization of pathogens from solid waste
- Contaminated drinking water

Pathogenic organisms can infect humans through contact and open skin wounds. This is especially relevant to workers in jobs related to solid waste and others exposed to solid waste, such as scavengers in dumps (see Figure 1.3).

Figure 1.3 shows an open dump I saw in Wuhan, China. These conditions are often prevalent in developing countries in Asia and Africa. Individuals coming in contact with trash or garbage can be infected through wounds. The individuals can bring pathogens on their clothing and shoes to the home and infect other individuals. Pathogens can also leach through the soil to water resources. Pathogenic bioaerosols can be released through the air and infect the respiratory system of individuals in the area.

The October 27, 2012, *New York Times* reported on the hazards from an open landfill or dump in Bangalore, India. It stated that: "A stinking mountain of trash, the landfill has been poisoning local waters and sickening nearby villagers. ... Trash is India's plague." The surface disposal of trash directly influences workers and scavengers. As stated in the article, as many as 15,000 waste workers can be affected.

In most developed countries, landfills require a liner to protect ground- and surface waters. Under these conditions, especially since the conditions in the landfill are under anaerobic conditions and produce methane, the survival of pathogens is negligible. The concern for water contamination is from inorganic and organic chemicals. There are few data on pathogen survival under lined landfill conditions in developed countries. These conditions are not usually applicable in developing countries.

Obire and Aguda (2002) in Nigeria found large numbers of total viable aerobic heterotrophic bacteria for the leachate from a waste dump and an adjacent stream. The leachate counts ranged from 2.5×10^6 to 6.5×10^6 CFU/mL, and in the stream counts ranged from 1.2×10^6 to 1.2×10^7 CFU/mL. The bacteria found in some leachate and stream samples were *Bacillus* spp., *Staphylococcus* spp., *Klebsiella*, and *Shigella* (Obire and Aguda 2002).

Flores-Tena et al. (2007) reported that they identified 20 pathogenic or opportunistic bacteria in air, 20 from soil, and 11 from leachate in a landfill in Mexico. Although most were enteric, some were respiratory tract pathogenic bacteria. In leachate, the most frequent species were *Acinetobacter baumannii*, *Bordetella* spp., *Brucella* spp., and *Escherichia coli* var II (Flores-Tena et al. 2007).

Aerosolization of pathogens can be from land application of sewage sludge and biosolids (Epstein 1998a, 1998b; Dowd et al. 2000). The extent of pathogen aerosolization depends on the type of application and weather conditions. Typically, biosolids (i.e., treated sewage sludge) in developed countries are either applied in a semisolid form directly to the land and followed by disking or other methods of incorporation into the soil, or liquid biosolids may be sprayed and then disked into the soil. Spraying will tend to disperse pathogens further than if semisolids are applied. Windy conditions could result in greater dispersion of bioaerosols (Epstein 1998). This is discussed further in Chapter 4. Bioaerosols may be dispersed during the composting of sewage sludge or biosolids. The aerated static pile (ASP) results in much fewer emissions of bioaerosols than the windrow method. In the windrow method, the turning and agitation could result in significant dispersion of bioaerosols containing pathogens. Enclosed systems with proper containment and treatment of the air will not result in emissions containing pathogenic bioaerosols (Epstein 1998b).

In many developing countries, fecal matter or raw sewage does not go through a sewage system and is often deposited on land. This could result in dispersion or bioaerosolization of pathogens or direct infection through contact.

In the case of incineration of sewage sludge or solid waste, the potential for pathogen infection is to workers. This could result from direct contact as well as inhalation of pathogenic bioaerosols. Workers need to practice good hygienic practices, use protective equipment, and leave clothing and shoes at the workplace.

SUMMARY

There are few data and information on pathogens in various solid wastes. The exposure to pathogens from solid wastes in developed countries is primarily a concern for workers. Respiratory and dermal modes of infection are probably the greatest in developing countries, especially in Asia and Africa, but also are important in Mexico and Central and South America.

Contamination of water supplies beneath dumps or in villages and small towns that do not have proper sewage infrastructure can result in diseases in the population.

Scavengers in dumps not only expose themselves through dermal and respiratory infections but also can bring causative elements into their homes on contaminated shoes, sandals, bare feet, and clothing, thus exposing other members of the family.

The main objective of this book is to discuss pathogens in solid waste, provide some of the recent available data, and provide insight into disposal options and management of solid waste to reduce infection and diseases.

REFERENCES

Abrahams PW. 2002. Soils: their implications in human health. *Science Total Environ* 291: 1–32.
American Academy of Allergy, Asthma, and Immunology. 2011. Asthma statistics. http://www.aaaai.org/about-the-aaaai/newsroom/asthma-statistics.aspx
Baumgardner DJ. 2012. Soil-related bacterial and fungal infections. *J Am Board Fam Med* 25: 734–744.
Berg G, Eberl L, and Hartman A. 2005. The rhizosphere as a reservoir for opportunistic human pathogenic bacteria. *Environ Microbiol* 7: 1673–1685.
Bolognia JL, Jorizzo JL, and Rapini RP (Eds.) 2007. *Dermatology* (2 volume set) 2nd edition. London: Elsevier.
Bowman DD. 2009. *Manure Pathogens*. Alexandria, VA: WEF Press.
Brooker S, Clements ACA, and Bundy DAP. 2006. Global epidemiology, ecology, and control soil-transmitted helminths infections. *Adv Parasitol* 62: 221–261.
Brundtland G. 1999. *Removing Obstacles to Health Development*. Geneva, Switzerland: World Health Organization.
Centers for Disease Control and Prevention (CDC). 2010. Diarrheal diseases in less developed countries. http://www.cdc.gov/healthywater/hygiene/ldc/diarrheal_diseases.html
Dowd SE, Gerba CP, Pepper IL, and Pillai SD. 2000. Bioaerosol transport modeling and risk assessment in relation to biosolid placement. *J Environ Qual* 29: 343–348.
Ebisawa I, Takayangi M, Kurata M, and Kigawa M. 1986. Density and distribution of *Clostridium tetani* in the soil. *Jpn J Exp Med* 56: 69–74.
Epstein E. 1998a. Legal and health implications of land application of biosolids. In *12th Annual Residuals and Biosolids Management Conference*. Bellevue, WA: Water Environment Federation, pp. 507–518.
Epstein E. 1998b. Pathogenic health aspects of land application. *BioCycle* 39: 62–67.
Flores-Tena FJ, Guerrero-Barrera AL, Alelar-Gondzalez FJ, Ramire-Lopez EM, and Martinez-Saldaria MC. 2007. Pathogenic and opportunistic Gram-negative bacteria in soil, leachate and air in San Nicolas landfill at Aguascalientes, Mexico. *Rev Latinoam Microbiol* 49: 25–30.
George R. 2008. *The Big Necessity*. New York: Metropolitan Books.
George G, and Ndip E. 2011. Prevalence of geophagia and its possible implications to health—a study in rural South Africa. Paper presented at 2011 2nd International Conference on Environmental Science and Development, Singapore.
Gilliss D, Cronquist A, Carter M, Tobin-D'Angelo M, Blythe D, Smith K, Lathrop S, Birkhead G, Cieslak P, Dunn J, Holt KG, Guzewich JJ, Henao Ol, Mahon B, Griffin P, Tauxe RV, and Crim SM. 2011. Vital signs: incidence and trends of infection with pathogens transmitted commonly through food. *MMWR Morb Mortal Wkly Rep* 60: 749–755.
How the other half dies. n.d. http://www.imva.org/Pages/deadtxt.htm
Krewski D, Balbus J, Butler-Jones D, Hass C, Isaac-Renton J, Roberts KJ, and Sinclair M. 2002. Managing health risk from drinking water—a report on the Walkerton inquiry. *J Toxicol Environ Health A* 65: 1635–1823.

MacKenzie WR, Hoxie NJ, Proctor ME, Gardus MS, Blair KA, Peterson DE, Kazmierczak JJ, Addiss DG, Fox KR, Rose JB, and Davis JP. 1994. A massive outbreak in Milwaukee of *Cryptosporidium* infection transmitted through the public water supply. *N Engl J Med* 331: 161–167.

Polymenakou PN, Manolis M, Euripides GS, and Anastasios T. 2008. Particle size distribution of airborne microorganisms and pathogens during an intense dust event in the eastern Mediterranean. *Environ Health Prospect.* 116: 292–296.

Ritter L, Solomon K, Sibley P, Hall K, Keen P, Mattu G, and Linton B. 2002. Sources, pathways, and relative risks of contaminants in surface water and ground water: a perspective prepared for the Walkerton inquiry. *J Toxicol Environ Health A* 65: 1–142.

Sing D and Sing CF. 2010. Impact of direct soil exposure from airborne dust and geophagy on human health. *Int J Environ Res Public Health* 7: 1205–1223.

Sobsey MD. 1978. Field survey of enteric viruses in solid waste landfill leachates. *Am J Public Health* 68: 858–864.

Teschke K, Bellack N, Shen H, Atwater J, Chu R, Koehoom M, MacNab YC, Schreier H, and Isaac-Renton JL. 2010. Water and sewage systems, socio-demographics, and duration of residence associated with endemic intestinal infectious diseases. *BMC Public Health* 10: 1–24.

Vermeer DE and Frate DA. 1979. Geophobia in rural Mississippi: environmental and cultural contents and nutritional implications. *Am J Clin Nutr* 32: 2129–2135.

World Health Organization (WHO). 2007. Fact Sheet No. 237. http://foodhygiene2010.files.wordpress.com/2010/06/who-food_safety_fact-sheet.pdf

World Health Organization (WHO). 2009. Diarrhoeal diseases. February 2009. http://www.who.int/vaccine_research/disease/diarrhoeal/en/index.html

2 Risk and Exposure

INTRODUCTION

When dealing with the potential for disease caused by pathogens, it is important to understand the risk potential. In this way, we can communicate to the individual and the public the precautions needed. It also allows public health individuals to evaluate and provide proper management. Risk analysis provides a framework for regulations.

There are virtually no risk assessments with respect to solid waste. For example, what is the risk to a young person rummaging through a dump to collect items that may be sold? Considering the prevalence of thousands of dumps in Asia and Africa as well as in South America, what is the risk of disease for workers, scavengers, and the public? How many persons have been infected?

Most of the risk assessments have been on food, drinking water, and chemicals. In 1993, the United States Environmental Protection Agency (USEPA) published Title 40 of the Code of Federal Regulations, Part 503. This set pollution limits for some organic and inorganic chemicals in the disposal of sewage sludge. Included also was a technical support document for reduction of pathogen and vector attraction. The risk assessment used by the USEPA in 1995 and updated subsequently follows four basic steps:

- Hazard identification: Can the identified pollutants harm human health and the environment?
- Exposure assessment: Who is exposed, how do they become exposed, and how much exposure occurs?
- Dose–response evaluation: What is the likelihood of an individual developing a particular disease as the dose and exposure increase?
- Risk characterization: What is the likelihood of an adverse effect in the population exposed to a pollutant? Risk is calculated as

$$\text{Risk} = \text{Hazard} \times \text{Exposure}$$

A major criticism of the USEPA regulations was the minimum emphasis and concern regarding exposure and potential disease from pathogens and pathogenic substances. One problem was the lack of dose–response relationships. Since that time, there have been attempts to improve the risk analysis, many of these through modeling since dose–response information was still lacking and is difficult to obtain.

It is not the objective of this chapter to go into extensive detail. However, when discussing pathogens, pathogenic substances, and diseases, some attention must be given to the risk potential. With respect to solid waste management, three populations are of major consideration:

- Workers handling or involved in the disposal of wastes
- Populations residing near waste disposal facilities
- Scavengers

In recent years, the risk paradigm called quantitative microbiological risk assessment (QMRA) has focused on exposure assessment, dose–response analysis, and risk characterization. The following are examples of three governmental organizations that have used QMRA:

- USEPA: *Cryptosporidium* risk and virus risk
- Food and Drug Administration (FDA): *Listeria monocytogenes*, *Escherichia coli*, *Vibrio parahaemolyticus*, all with regard to certain foods
- United States Department of Agriculture: *Clostridium perfringens*, *Salmonella* serotype Enteritidis and *E. coli* O157:H7, also with regard to foods

What is risk? Risk is the probability (i.e., the likelihood) of exposure and the result that something dangerous or hazardous can occur related to human health. In the case of potential diseases to humans resulting from the potential exposure to pathogens, it is important to determine the conditions of exposure, the pathogens of concern, routes of exposure, and the risk factors. This involves a risk analysis.

The three basic components of risk analysis are

- Risk assessment
- Risk communication
- Risk management

These are interrelated. However, unless one conducts or evaluates the risk assessment, it is difficult to manage an unknown and therefore to communicate the consequences and how to proceed in managing and controlling the potential for disease.

Although this chapter is a summary and guideline, there are several excellent sources for those who wish to become further involved in the subject. Some references are provided in the Suggested Reading section at the end of the chapter.

Haas (1999) presented an early framework for risk analysis illustrated in Figure 2.1, which shows the integrated components of the risk.

In 1995, the USEPA updated and issued the current Agency-wide Risk Characterization Policy (USEPA 1995). The policy calls for all risk assessments performed at the USEPA to include a risk characterization to ensure that the risk assessment process is transparent; it also emphasizes that risk assessments are to be clear, reasonable, and consistent with other risk assessments of similar scope prepared by programs across the agency. Effective risk characterization is achieved through transparency in the risk assessment process and clarity, consistency, and reasonableness of the risk assessment product. Risk characterization, the last step in risk

FIGURE 2.1 Risk analysis framework. (From Haas, 1999. *Quantitative Microbial Risk Assessment*. New York: Wiley.)

assessment, is the starting point for risk management considerations and the foundation for regulatory decision making, but it is only one of several important components in such decisions. As the last step in risk assessment, the risk characterization identifies and highlights the noteworthy risk conclusions and related uncertainties. Each of the environmental laws administered by the USEPA calls for consideration of other factors at various stages in the regulatory process. As authorized by different statutes, decision makers evaluate technical feasibility (e.g., treatability, detection limits) and economic, social, political, and legal factors as part of the analysis of whether to regulate and, if so, to what extent. Thus, regulatory decisions are usually based on a combination of the technical analysis used to develop the risk assessment and information from other fields.

Recognizing that for many people the term *risk assessment* has wide meaning, the National Research Council (NRC) report in 1983 on risk assessment in the federal government distinguished between risk assessment and risk management: "Broader uses of the term [risk assessment] than ours also embrace analysis of perceived risks, comparisons of risks associated with different regulatory strategies, and occasionally analysis of the economic and social implications of regulatory decisions functions that we assign to risk management."

In 1984, the USEPA endorsed these distinctions between risk assessment and risk management for agency use and later relied on them in developing risk assessment guidelines. In 1994, the NRC reviewed the agency's approach to and use of risk assessment and issued an extensive report on the findings. This distinction suggests that USEPA participants in the process can be grouped into two main categories, each with somewhat different responsibilities, based on their roles with respect to risk assessment and risk management (USEPA 1995).

The NRC developed a standardized model for the purpose of developing a uniform risk assessment methodology for all federal agencies. The NRC model has four steps: (1) risk assessment, (2) hazard assessment and identification, (3) dose–response assessment, and (4) exposure assessment.

RISK ASSESSMENT

Modern societies have learned to reduce the impact of disease-causing microorganisms (pathogens) by adopting various sanitary control measures, such as farm-to-fork processes in food production and treatment plants for drinking water and wastewater. Nonetheless, our aging and more vulnerable population groups combined with the emergence of drug-resistant pathogens and enhanced global spread of human pathogens provide a breeding ground for novel and reemerging diseases. What is the need for risk assessment? Whenever a hazard exists, such as potential exposure to a pathogen or pathogens in this case, it is valuable to assess the potential risk. The greater the potential risk and exposure, the more precautions are necessary. As indicated, in most cases workers are the most exposed individuals. They are exposed more frequently and often to higher amounts of the hazard.

The risk assessment process is often provided as a four-step process:

- Hazard assessment and identification: This process identifies the pathogens and diseases caused by them that could result in infection morbidity and mortality.
- Exposure assessment: This delineates the likelihood of an individual developing the disease as the dose and exposure increase.
- Dose–response analysis. The relationship between the hazard quantity and the adverse effect is explored.
- Risk characteristics: This synthesizes the previous steps to estimate the risk.

Most risk analysis has been geared to chemical hazards since these are easier to evaluate. Furthermore, the potential of incurring a disease or infection is often related to the number of pathogens exposed and the occurrence of exposure. For example, dermatological exposure when there are no eruptions in the skin or fissures that allow the pathogen to enter is less apt to result in the incurrence of a disease.

Risk assessment allows for planning, hazard identification, exposure mediation, and prevention. Risk assessment also provides for the identification of the most likely population that could be affected. Generally, children and aged populations are at greater risk.

Parkin (2008) comprehensively reviewed the foundations and frameworks for human microbial risk assessment. The author pointed out that the early concept of risk assessment was provided by the NRC as follows: The framework is based on the 1983 NRC concept. Although this model was based on chemical risks, it was applied to pathogens. Subsequently, in the 1990s, additional issues were identified, and it was determined that the model based on chemical risk was not applicable to microbial pathogens (Haas et al. 1993). The microbial risk assessment (MRA) is fundamentally different from chemical risk assessment (Eisenberg 2006). Eisenberg indicated that the chemical risk assessment to a great extent does not account for infectious diseases and immunity issues. Furthermore, the chemical risk assessment relies on static modeling techniques, which cannot represent dynamic processes such as disease transmission (Eisenberg 2006; Parkin 2008). The traditional risk assessment is shown in Figure 2.2.

Risk and Exposure

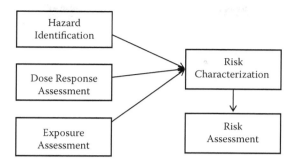

FIGURE 2.2 Traditional risk assessment framework.

Several problems with using a chemical risk analysis for pathogens or microorganisms were identified by scientists (Parkin 2008):

- Pathogens can grow, evolve into different life stages, and die off.
- Virulence, the relative power and degree of pathogenicity by organisms, varies during a pathogen's life cycle and between strains.
- Pathogens behave differently under different ecological and environmental conditions, such as temperature, moisture, and media (soil).
- Pathogens may be transmitted from person to person. Chemicals are not.
- Infection and exposure differs between individuals. Immunities vary among persons.

Subsequently, an ecological risk assessment framework was developed in North America (USEPA 1998; Environment Canada 1996).

HAZARD ASSESSMENT AND IDENTIFICATION

Hazard assessment relates to identification of a pathogen of significance that may cause acute or chronic illnesses in humans. Pathogenic microorganisms are essentially known by the scientific community. Pathogens are often referred to as primary pathogens and secondary or opportunistic pathogens. The primary pathogens can be bacteria, parasites, or viruses. They usually involve direct contact. With solid waste, workers are the individuals most exposed. Good hygiene is the best preventive measure. Epidemiological or ecological studies, including exposure routes and the potential effects on individuals, provide information on the distribution of disease and pathogens among persons.

The two important exposure routes are through either direct contact or bioaerosols. Direct routes could be ingestion, inhalation, or dermal contact. Ingestion can come from contaminated food or water. Inhalation occurs from air pollution or emissions of bioaerosols. Dermal contact is through direct contact with pathogen-contaminated material. Direct contact typically involves individuals. A good example is the fecal-to-oral route.

In developing countries where sanitation and hygiene are poor, many of the diseases causing morbidity and mortality occur this way. Often, in rural areas of developing countries, fecal defecation occurs outdoors, where children in particular are exposed. Exposure can occur also when persons, especially children, come in contact with solid waste either at home or in public places such as dumps. In developed countries, exposure is broader and generally is of concern to the public as a whole. This could be from air pollution from incineration or bioaerosol emissions from facilities such as those involved with solid waste composting.

Another exposure route could be from groundwater pollution as a result of old landfills that do not have liners or from animal waste, particularly in areas where large concentrations of domestic animals are outdoors. In dealing with solid waste exposure, consideration must be given to (1) workers and (2) the public. Worker exposure can be either through direct contact or through bioaerosols. The public will be exposed primarily through pathogenic bioaerosols.

DOSE–RESPONSE ASSESSMENT

Dose–response assessment is the process of quantitatively evaluating the health aspect of a specific disease as a function of human exposure to that pathogen. The relationship between the dose of the organism exposure and the incidence of adverse health effects in the exposed population forms the basis for the quantitative dose–response relationship. The response of humans to exposure to a pathogen is highly variable. The incidence of disease is dependent on a number of factors, such as virulence characteristics of the pathogen, the number of cells ingested, and the general health and immune conditions of the hosts. Generally, children and older persons are more susceptible to a disease from a pathogen. Immunosuppressed individuals are also more susceptible. These last individuals are much more susceptible to an infection (e.g., from bioaerosols) than the general population. A key question when comparing dose–response studies is which biological responses are being measured. In the case of pathogenic enteric bacterial infection, morbidity and mortality are the end points (Buchanan, Smith, and Long 2000). The term *infections* is often defined differently by various disciplines.

Any consideration of microbial dose–response relations must take into account the various modes of pathogenicity associated with different pathogenic organisms. An understanding of how a pathogen causes a disease is critical. Three different broad classes of pathogens are differentiated—infectious, toxico-infectious, and toxigenic—based on the modes of pathogenicity (Buchanan, Smith, and Long 2000).

EXPOSURE ASSESSMENT

Exposure is the frequency, duration, and intensity with which a population or persons are exposed through various routes, such as inhalation, ingestion, or dermal contact. Inhalation with regard to solid waste exposure can be through bioaerosols. Ingestion usually involves contaminated food or water. The exposure assessment process includes the following steps: (1) characterize exposure setting, (2) identify exposure pathways, and (3) quantify exposure.

RISK CHARACTERIZATION

Risk characterization is the final phase of risk assessment. During this phase, the likelihood of adverse effects occurring as a result of exposure to a stressor is evaluated. Risk characterization contains two major steps: risk estimation and risk description. The stressor–response profile and the exposure profile from the analysis phase serve as input to risk estimation. The uncertainties identified during all phases of the risk assessment also are analyzed and summarized. The estimated risks are discussed by considering the types and magnitude of effects anticipated, the spatial and temporal extent of the effects, and recovery potential. Supporting information in the form of a weight-of-evidence discussion also is presented during this step. The results of the risk assessment, including the relevance of the identified risks to the original goals of the risk assessment, then are discussed with the risk manager (EPA/630/R-92/001 Framework for Ecological Risk Assessment).

REFERENCES

Buchanan RL, Smith JI, and Long W. 2000. Microbial risk assessment: dose-response relations and risk characterization. *Int J Food Microbiol* 58: 159–172.

Eisenberg JNS. 2006. *Application of a Dynamic Model to Assess Microbial Health Risks Associated with the Beneficial Uses of Biosolids: Final Report to the Water Environmental Research Foundation.* WERF Report 98-REM-1a. London: Water Environmental Research Foundation.

Environment Canada. 1996. *Ecological Risk Assessment of Priority Substances under the Canadian Protection Act.* Ottawa: Environment Canada.

Haas, C. 1999. *Quantitative Microbial Risk Assessment.* New York: Wiley.

Haas C, and Eisenberg JNS. 2001. Risk assessment. In *Water Quality: Guidelines, Standards and Health*, ed. Bartram L, and Fewtreil J. London: World Health Organization.

Haas CN, Rose JB, Gerba C, and Regli S. 1993. Risk assessment of virus in drinking water. *Risk Anal* 13: 545–662.

National Research Council (NRC). 1983. *Risk Assessment in the Federal Government: Managing the Process.* Washington, DC: National Academy Press.

Parkin RT. 2008. *Foundations and Frameworks for Human Microbial Risk Assessment.* Washington, DC: US Environmental Protection Agency.

US Environmental Protecton Agency (USEPA). 1993. Title 40, Code of Federal Regulations, Part 503.

US Environmental Protection Agency (USEPA). 1995. *Guidance for Risk Characterization.* Washington, DC: US Environmental Protection Agency.

US Environmental Protection Agency (USEPA). 1982, November. *Technical Support Document for Reduction of Pathogens and Vector Attraction in Sewage Sludge.* Washington, DC: USDA.

US Environmental Protection Agency (USEPA). 1995. *A Guide to the Biosolids Risk Assessments for the EPA Part 503 Rule. US Environmental Protection Agency.* Washington, DC: USDA.

US Environmental Protection Agency (USEPA). 1998. *Guidelines for Ecological Risk Assessment*. Washington, DC.: *Federal Register* 63(93): 26846–26924.
United States Environmental Protection Agency and US Department of Agriculture (USEPA/ USDA). 2012. *Microbial Risk Assessment Guideline*. Washington, DC: United States Environmental Protection Agency and US Department of Agriculture.

SUGGESTED READING

Haas CN, Rose JB, and Gerba CP. 1999. *Quantitative Microbial Risk Assessment*. New York: Wiley.
Parkin RT. 2008. *Foundations and Frameworks for Human Microbial Risk Assessment*. Report submitted to the US Environmental Protection Agency, Washington, DC.
US Environmental Protection Agency (USEPA) and US Department of Agriculture. 2012. *Microbial Risk Assessment Guideline. Pathogenic Microorganisms with Focus on Food and Water*. Washington, DC: Interagency Microbiological Risk Assessment Guideline Workgroup.
World Health Organization (WHO). 2001. *Water Quality, Guidelines, Standards and Health*, ed. Fewtrell L, and Bartram J. London: IWA.

3 Odors as a Health Issue

INTRODUCTION

Odors are pervasive throughout the globe. Primarily, odors are either chemical or organic. Typically, chemical odors are the result of industrial activity, including from

- Pulp and paper mills
- Wood treatment plants
- Petroleum facilities
- Pesticide and fertilizer plants
- Incineration facilities

Organic odors usually result from

- Wastewater treatment plants
- Septage and sewage
- Composting facilities
- Landfills and open dumps
- Land application of wastes
- Animal farms, especially large concentrations of animals
- Human and animal discharge to fields, streets, and open areas
- Garbage discharged in streets

BASIC CONCEPTS

What makes some of the odors from solid waste? Odorants result from the decomposition of organic matter, primarily containing sulfur and nitrogen. What is the difference between an odor and an odorant? The term *odor* refers to the perception experienced when one or more chemicals come in contact with receptors on olfactory nerves. An *odorant* refers to any chemical in the air that is part of the perception of odor (McGinley and McGinley 1999).

How are odors characterized?

- *Odor quality*. Odor levels are usually expressed as the ratio of dilution to threshold (D/T) rather than concentration. One of the major reasons for not characterizing the odor by concentration of specific compounds is that one compound at low concentration can be much more effective in producing a malodor than another compound at a high concentration.

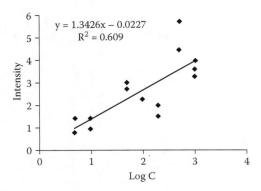

FIGURE 3.1 Intensity versus concentration of dimethyl sulfide.

- *Odor intensity.* This is the relative strength of the odor compared to a standard compound, usually butanol. It is expressed in parts per million (ppm) of butanol. This is expressed by Steven's law. The equation is $I = kC^n$ where I is intensity, C is the mass concentration of the odorant I (mg/m^3) (Epstein 2011; Hooper and Cha 1988; Walker 1993; Water Environment Federation [WEF] 2004). An example is shown in Figure 3.1. As the concentration (log C) increases, the intensity increases.
- *Odor persistence.* This indicates how long the odor prevails and is an indication of the rate of dilution.
- *Odor character.* This is the type of odor or its offensiveness. For example, hydrogen sulfide smells like rotten eggs, and dimethyl sulfide smells like rotten cabbage.
- *Hedonic tone.* The hedonic tone is the relative pleasantness or unpleasantness of an odor.

Principally, most odors are considered a nuisance. However, in the 1990s scientific interest began to evaluate whether odors per se could result in health effects. Until that time, there were few scientific papers indicating that odors may be a health problem. In the United States, individual states regulated odors in a vague way. Most states did not use quantifying or mathematical ways to evaluate odors.

The assessment of potential health effects as a result of odors is difficult. Some of the reasons for these difficulties are the following:

- A small quantity of an odorant produced by a chemical well below a toxic level can result in an obnoxious odor.
- Odor is generally considered a nuisance or as a damaging effect on the quality of life.
- Perception of an odor as noxious is not directly linked to toxicity.
- Behavioral patterns and physiological responses can be developed as a result of exposure to odors.
- Epidemiological studies evaluating the effect of odors on health are limited.

Odors as a Health Issue

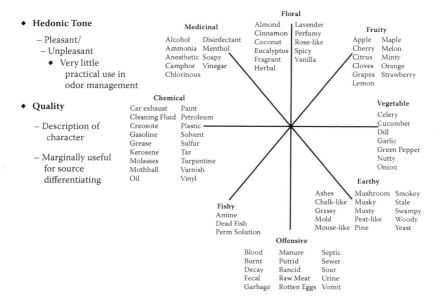

FIGURE 3.2 The quality and hedonic tone of odors.

- There are no toxicity values for use in assessing risk posed by odorants commonly present in biosolids.
- There is no database to properly assess odorants and associated potential risks.

Figure 3.2 shows odor quality and hedonic tone.

Once an odor is detected, its effect and consequences increase with concentration as shown in Table 3.1. Figure 3.3 shows the effect of odor concentration on eye and nasal irritation.

TABLE 3.1
Sequence of Sensory Effects as a Result of an Odorant Concentration

Concentration	Level	Effect
	1	Odor detection
	2	Odor recognition
	3	Odor annoyance
	4	Odor intolerance
	5	Perceived
	6	Intolerance
	7	Toxicity

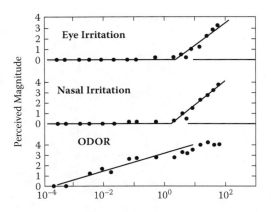

FIGURE 3.3 The effect of odor concentration on eye and nasal irritation.

An odor episode can lead to complaints from the public and is related to (1) odor character, (2) odor intensity, (3) odor duration, and (4) odor frequency (McGinley and McGinley 1999). The sensitivity and response to odors will vary in individuals. This variation can be related to age, gender, social habits, and medical history. The medical conditions that can increase sensitivity to odors are asthma, chronic obstructive pulmonary diseases (COPDs), mental health (e.g., depression), and hypersensitivity.

As we inhale air containing an odor, the following occurs:

- Ten percent of odors pass under the olfactory organ.
- Twenty percent pass under the epithelium during sniffing. There are 10 to 25 million olfactory cells in the epithelium.
- The mucous layer on the epithelium traps odorants that are water soluble.
- An electrical response is created that, depending on its amplitude (strength), is sent along to the brain in the form of a pain stimulus.

The sense of smell is complex and unique in structure and organization. The human olfactory system supplies 80% of flavor sensation during eating. However, the olfactory system plays a major role as a defense mechanism by creating an aversion response to malodors and irritants (McGinley and McGinley 1999). The odorant receptors and the organization of the olfactory system are shown in Figure 3.4.

Odors are experienced differently by different people, such as whether the perfume of another individual is bothersome to someone in an elevator, yet the individual and others do not object to the odor. Odor can trigger memories and associations.

There are numerous factors affecting odor perception. These can include adaptation, environmental exposure, age, nutritional status, mental illness, diseases, and disorders (Chrostowski and Foster 2003). Dalton et al. (2002) indicates that there are numerous factors, including exposure history, personality, beliefs, expectations, social factors, and bias. The longer one is exposed to an odor, especially if its intensity increases or it extends for long periods of time, the more apt one is to consider

Odors as a Health Issue

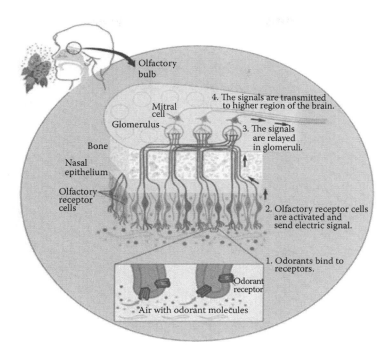

FIGURE 3.4 Odorant receptors and the organization of the olfactory system. (From Cain WS, presentation at US Composting Council's 14th Annual Conference and Travel Show, Albuquerque, NM, 2006, with permission.)

whether the odor can have a health effect. Perception can be an important factor. In a study when subjects were told that an odor was hazardous, the odor was more offensive than when told it was a natural occurrence.

The following odorants detected in biosolids and associated with wastewater treatment were reported by the National Resources Council (NRC 2002).

Sulfur compounds
- Hydrogen sulfide
- Dimethyl sulfide
- Diphenyl sulfide
- Carbon disulfide
- Dimethyl disulfide
- Methyl mercaptan
- Ethyl mercaptan
- Propyl mercaptan
- Allyl mercaptan
- Benzyl mercaptan
- Thiocresol
- Diethyl sulfide

- Diallyl sulfide
- Butyl mercaptan
- T-Butyl mercaptan
- Crotyl mercaptan
- Thiophenol
- Sulfur dioxide

Nitrogen compounds
- Ammonia
- Methylamine
- Dimethylamine
- Trimethylamine
- Ethylamine
- Triethylamine
- Pyridine
- Indole
- Skatole

Acids
- Acetic (ethanoic)
- Butyric (butanoic)

Aldehydes
- Acetaldehyde
- Diamines (cadaverine)

Aldehydes and ketones
- Butyraldehyde
- Formaldehyde
- Isobutyraldehyde
- Isovaleraldehyde
- Acetone
- Butanone

ODOR AND HEALTH

In my previous books, I discussed odor as related to health and provided extensive discussions on odor management (Epstein 1997, 2011). In this book, I primarily want to make information on the subject current. Excellent reviews were presented by several authors (Chrostowski and Foster 2003; Johnson 2011). Readers should also evaluate other references (Cain and Cometto-Muniz 2004; Epstein 2011; Schiffman and Williams 2005; Shusterman 1992).

Malodors may have negative effects on well-being. Convened on April 15–17, 1998 to discuss odors and health, the US Environmental Protection Agency (USEPA) and the National Institute of Deafness and Other Communication Disorders raised greater awareness, and interest evolved on whether odor has associated health effects (Schiffman and Williams 2005).

Odors as a Health Issue

Cain (1987) indicated that people associated the health or harmful effects of environments by the presence or quality of perceived odors.

Dalton (1999) reported that many health-related effects of exposure to odorants are mediated not by the direct effect of odors but by cognitive association of odors and health. In a study, Dalton found that individuals given a harmful bias reported significantly more health symptoms on exposure to an odorant than those receiving the same odorant but with no harmful effects indicated. Thus, the author concluded that prejudiced odor perceptions and reactions underscore the incredible ambiguity of odor sensation and suggested that similar nonsensory factors play a large role in people's everyday reactions to ambient odors. When odors are persistent, they can result in potential health effects (Dalton 1999).

Furthermore, odor perception has been shown to affect mood, tension, stress, depression, anger, and fatigue. These conditions could potentially lead to physiological and biochemical changes with subsequent health effects (Bolla-Wilson, Wilson, and Bleeker 1988)

We can view health effects in various ways. Depending on the frequency and extent of an odor, if one resides near an industrial source of odors, people complain of headaches, nausea, stress, eye irritation, throat irritation, cough, and mood changes. Today, these are considered health effects depending on the severity. Odor at times has served as an exposure marker.

Figure 3.4 shows that as the odor level increases, it eventually triggers an irritation. If the odor persists, this irritation can be great and require medical treatment.

It is evident from Table 3.2 that some compounds can cause an irritating effect at low concentrations, whereas others require much higher concentrations.

Villemure et al. (2003) reported that odors altered mood, anxiety level, and pain unpleasantness but did not change the perception of pain intensity (Villemure, Slotnick, and Bushnell 2003). Johnson (2011) found that odor delivery can have both a positive and a negative impact on cognitive operations (Johnson 2011).

TABLE 3.2
Odorous Compounds and Toxicological Thresholds of Irritation

Chemical	Odor Threshold Range (mg/m^3)	Irritating Concentration (mg/m^3)
Acetic acid	2.5–250	25
Acetone	47–1613	475
Acetaldehyde	0.0002–4.1	90
Ammonia	0.026–40	72
Hydrogen sulfide	0.0007–0.014	14
Methyl ethyl ketone	0.74–147	590

Source: Modified from Chrostowski PC, and Foster MS. 2003. Odor perception and health effects. WEFTEC 2003 Workshop on the Status of Biosolids Recycling in the United States, Los Angeles, October.

SUMMARY

The term *odor* refers to the perception experienced when one or more chemicals come in contact with receptors on olfactory nerves. An *odorant* refers to any chemical in the air that is part of the perception of odor. While odors are generally perceived as nuisances, if they persist or intensify, they can be an irritant, cause headaches, and especially affect people with respiratory conditions such as asthma, emphysema, and COPD. These are health concerns.

REFERENCES

Bolla-Wilson K, Wilson RJ, and Bleeker ML. 1988. Conditions of physical symptoms after neurotoxic exposure. *J Occup Med* 30: 684–686.
Cain WS. 1987. Indoor air as a source of annoyance. In *Environmental Annoyance, Characterization, Measurement and Control*, ed. Koelega HS. Amsterdam: Elsevier.
Cain WS and Cometto-Muiz JE. 2004. *Health Effects of Biosolid Odors: A Review*. Alexandria, VA: Water Environment Research Foundation.
Chrostowski PC, and Foster MS. 2003. Odor perception and health effects. Presented at the WEFTEC 2003 Workshop on the Status of Biosolids Recycling in the United States, Los Angeles, October.
Dalton P. 1999. Cognitive influence on health symptoms from acute chemical exposure. *Health Physiol.* 18: 579–590.
Dalton P et al. 2002. Gender-specific induction of enhanced sensitivity to odors. *Nat Neurosci* 5(3): 199–200.
Epstein E. 1997. *The Science of Composting*. Lancaster, PA: Technomic.
Epstein E, ed. 2011. *Industrial Composting: Environmental Engineering and Facilities Management*. Boca Raton, FL: CRC Press.
Hooper JE and Cha S. 1988. *Odor Perception and Its Measurement*. East Hartford, CT: TRC Environmental Consultants.
Johnson AJ. 2011. Cognitive facilitation following intentional odor exposure. *Sensors* 11: 5469–5488.
McGinley MA and McGinley CM. 1999. The "gray line" between odor nuisance and health effects. Proceedings of Air and Waste Management Association, 92nd Annual Meeting and Exhibition, St. Louis, MO.
National Research Council, July 2002. *Biosolids Applied to Land: Advancing Standards and Practices*. Washington, DC: National Academies Press.
Schiffman SS, Walker JM, Dalton P, Lorig TS, Rajmer JH, Shusterman D, and Williams CM. 2000. Potential health effects of odors from animal operations, wastewater treatment, and recycling of byproducts. *J Agromed* 7: 7–81.
Schiffman SS and Williams CM. 2005. Science of odor as a potential health issue. *J Environ Quality* 34: 129–138.
Shusterman D. 1992. Critical review: the health significance of environmental odor pollution. *Arch Environ Health* 47: 76–87.
Villemure CS, Slotnick BM, and Bushnell MC. 2003. Effects of odor on pain perception: deciphering the role of emotion and attention. *Pain* 106: 101–108.
Walker JM. 1993. Control of composting odors. In Hoitink HAJ and Keener HM, eds. *Science and Engineering of Composting: Design, Environmental, Microbiological, and Utilization Aspects*. Columbus, OH: Ohio State University.
Water Environment Federation (WEF). 2004. *Control of Odors and Emissions from Wastewater Treatment Plants*. WEF Manual of Practice 25. Alexandria, VA.: Water Environment Federation.

4 Pathogens and Diseases of Solid Wastes
Municipal Solid Waste or Garbage

INTRODUCTION

Unfortunately, little data exist on pathogens in municipal solid waste (MSW), dumps, and the handling of garbage, although the World Health Organization (WHO) does provide data and information available on the Internet. By reviewing the method of how MSW and garbage is handled globally, the lack of data is apparent. In 1983, Althaus et al. (Althaus, Sauerwald, and Schrammeck 1983; Collins and Kennedy 1992) examined wastes from hospitals and municipal garbage dumps and found that household wastes often contained more pathogens than hospital wastes.

In the United States, MSW is typically either landfilled or incinerated. Data from the United States Environmental Protection Agency (USEPA) are shown in Figures 4.1 and 4.2. In 2010, approximately 250 million tons of MSW were produced. Since 1960, MSW generation increased. In 2010, the generation rates were reduced. Generation per person remained at the same level from 1990 to 2010. The percentage of generated recycled material increased since 1985 to an average of 25%. In some states, it is much higher. There are several reasons for these trends. Many communities mandate recycling with a minimum separation of paper from glass and plastics and do not allow disposal of yard waste in landfills. The use of garbage disposals in the home reduces food waste removal. For many years, New York City prohibited garbage disposals in homes. More recently, this restriction was lifted, resulting in more liquid and greater biological oxygen demand (BOD) going into wastewater treatment plants and production of more sewage sludge. But, it preserved landfill space at the Staten Island landfill.

In the United States, most MSW goes to landfills. In recent years, regulations required that landfills be lined, and often the methane gas produced is collected and used. Earlier landfills were not lined; therefore, leachate could move through the media and contaminate groundwater or drinking water. How long the liners will remain intact is a controversial question.

Incineration is also a major method of handling MSW. However, in the United States, incineration is not as prominent as in Europe. Land is more available in the United States. Furthermore, many communities oppose incineration since they are

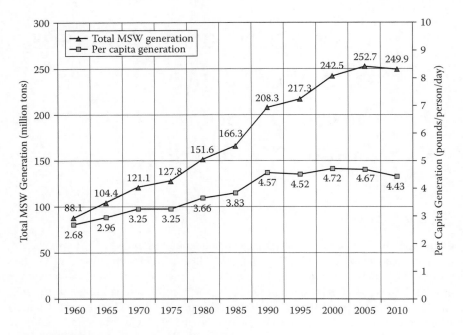

FIGURE 4.1 MSW generation rates in the United States.

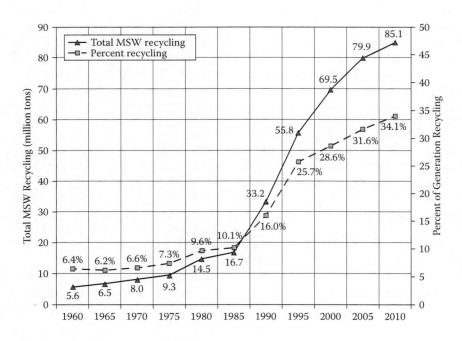

FIGURE 4.2 MSW recycling rates in the United States.

concerned with air pollution. It is difficult to permit an incinerator even if it uses heat recovery.

More recently in the United States and Canada, interest has been raised and effort has been made to recover energy. Many states and municipalities conduct recycling of paper, plastics, and glass. Cardboard is often recycled to Asia for reuse. Paper was recycled for production of pulp and new recycled paper. Glass is recycled, and attempts are also made to use crushed glass as a road surface material. The food industry recognized that placing food in dark glass containers was better for preservation. However, the public prefers to see the product; hence, clear glass is used.

The separation of glass into clear and colored elements is more labor intensive. But overall, separation techniques are improving. Metal recovery occurs through magnets for ferrous metals, and aluminum is collected using eddy current techniques. Air separation is used for lightweight material.

Aluminum, ferrous metals, plastics, and paper are all separated in a facility that I designed in Hong Kong. This was more expensive than landfilling the waste. Some grocery stores require consumers to bring recycled bags to reduce the distribution of plastic. Several grocery stores in the United States give consumers the option of plastic or paper bags. A few communities are planning or currently avoiding the use of plastic bags. In Hong Kong, Tuesday was a day free of plastic containers.

Attempts at composting or anaerobic digestion met with little success in the United States. In Europe, anaerobic digestion of source-separated MSW is used a bit more. In the United States and Canada, there is considerable composting of processed sewage sludge, termed biosolids. There is considerable composting of yard material since many states ban it from going into landfills. Recently, there is an increase in composting of food waste, primarily from large generators.

In Europe, land is at a premium. There is more emphasis on waste reduction, reuse, and recycling. For whatever cannot be recycled or reused, incineration is the optimal disposal method.

The public is most unlikely to be exposed directly to bioaerosols, diseases, and illnesses from MSW that is disposed of in landfills or incinerators.

Considering this situation in the United States, there is little interest in investigating the role of pathogens and their diseases in MSW. It is most likely that worker health is of greater importance. Good hygienic practices are important. Smoking or eating while working without disinfection can result in gastrointestinal diseases. Inhalation of bioaerosols is probably of greater importance than ingestion as the bioaerosols will infect the respiratory system.

The situation in developing countries is very different. The disposal of solid waste is haphazard and often ends in dumps, where scavengers attempt to find marketable material. In these cases, not only workers collecting the waste but also scavengers and their families are exposed.

In 1976, at the request of the United Nations Development Programs (UNDP), I went to Aden in Yemen to investigate the feasibility of composting solid waste by a German technology to be financed by the UNDP. The facility was to be designed by the United Nations Industrial Developing Organization (UNIDO). Although Aden was the most developed city in Yemen as a result of the previous presence of the

British, its infrastructure was primarily for wastewater. The German technology was not recommended because Yemeni personnel were not trained to maintain such a facility once the foreign professionals left. It was pointed out to the South Yemen government that similar plants in Morocco and Egypt were not in use and were not constructed correctly. It turned out that even these countries, which were much more advanced than Yemen, were not capable of maintaining the facilities.

The issue of collection and disposal is another solid waste problem I have seen in the Middle East and Asian countries. Garbage often left in streets for a considerable amount of time exposes children and other residents to pathogens and pathogenic bioaerosols. In Chapter 1, a dump in Wuhan, China, shown in Figure 1.3, is exemplary of conditions prevalent throughout Asia, Africa, Mexico, and other Central and South American countries. Workers, scavengers, and the public are at risk of exposure and conditions present a threat to public health, especially since hygiene is not well practiced and disinfectants are scarcely available.

In Egypt, a country of 80 million persons, 75% of the solid waste is generated in urban areas. It is estimated that by 2025, MSW will reach 33 million tons per year. Eight percent of the solid waste is sent to a compost plant, and the rest is disposed in dumps in open spaces, presenting a major health risk to the public (Khatib 2011; Collins and Kennedy 1992). Similarly, this situation exists in many Asian and African countries.

What are the health impacts of solid waste? Unless MSW is properly disposed, when discarded, especially for organic wastes, decomposition occurs that creates favorable conditions for the survival and growth of pathogens. Wastes harboring pathogens can be diapers, clothes containing fecal matter, blood, animal flesh, body fluids, soiled rags, and wastes that attract flies, rats, and other vectors and creatures. The most vulnerable persons are handlers and workers, as well as children and pickers frequenting dumps.

When dumps are located adjacent to bodies of water that serve as a source of drinking water, the additional potential from water contaminated with pathogens exposes the local population. Frequently, dumps or waste disposal sites are located upstream of wells and sources of drinking water.

The occupational hazards related to disposal and handling of solid waste are as follows:

- Infections
  - Skin and blood infections obtained from direct contact with wastes and open wounds
  - Eye infections from exposure to dust
  - Respiratory infections from exposure to pathogenic bioaerosols
  - Infections and diseases passed on through vectors
- Chronic diseases
  - Infections from medical wastes discharged as part of solid waste
  - Exposure from industrial discharge of chemicals (not covered in this book)
  - Accidents, such as infected wounds from exposure of ruptured skin or burns from methane gas explosions at landfill sites

Pathogens and Diseases of Solid Wastes

FIGURE 4.3 Scavengers at a dump in Wuhan, China.

Health risks from waste are caused by numerous factors, including the following:

- The composition and type of raw waste
- The decomposition of the waste and its release, leading to the environmental potential for diseases or infection (e.g., dust, attraction to vectors)
- Occupational risks related to handling and processing of the waste
- Disposal of wastes

Dumps in developing countries are a source of recyclable material. In developed countries, recycling usually takes place at the source or at a mechanized recycling center. An example is Palm Beach County, Florida. In dumps in developing countries, often children and women are involved in rummaging through the material, hence creating a greater risk of exposure. Furthermore, the lack of proper hygiene increases the exposure of the individuals and their families. Most likely, workers in dumps such as shown in Figure 4.3 do not change clothes when arriving home or traveling. I even saw a child in a baby carriage waiting while the mother scavenged the dump. Both the workers and the scavengers are contaminated directly or through bioaerosols.

PATHOGENS AND THEIR DISEASES

A pathogen is an organism capable of causing disease. It can be a virus, bacteria, microorganisms, fungi, or actinomycetes. Often, they are classified as primary and secondary pathogens. Primary pathogens can invade and infect a healthy person. The young and aged are the most vulnerable. Secondary pathogens primarily invade and infect a debilitated person or an individual on immunosuppressant medication. Good examples are diabetics, especially those with type I diabetes, but those with type II can also be affected. The immune system of these individuals is depressed, and they are more subject to infections. How many organisms are needed to cause an infection was addressed in the discussion of risk assessment, especially dose–response relationships or infective dose.

The significant aspect of diseases resulting from poor sanitation is illustrated by the bubonic or pneumonic plague in 1994 in Surat, India. Several hundred thousand people fled the city. This represented 25% of the population of about 1.5 million. Approximately 6,000 persons became ill, and 56 died. The cost to the economy of India was estimated at $600 million. Vectors were a significant factor in the spread of the disease.

As indicated, the leachate from landfills in developed countries currently is contained by liners. There is also a requirement to cover the waste disposed daily. This prevents the spread of diseases, especially pathogenic bioaerosols. Previously, landfills were essentially soil-covered dumps. The leachate may enter groundwater, a source of drinking water. Engelbrecht and Amihor (1975) found several pathogens in landfill leachate: *Staphylococcus aureus, Streptococcus pyogenes, Streptococcus faecalis, Streptococcus durans, Streptococcus pneumonia, Klebsiella pneumoniae,* various S*almonella* serovars, as well as *Proteus* spp. and coliform bacilli.

Pourcher et al. (2001) identified aerobic cellulolytic bacteria in both 1- and 5-year-old refuse samples from a French landfill site. The organisms ranged from 1.1×10^6 to 2.3×10^8 CFU (colony-forming units) (g.dry wt)$^{-1}$. The predominant groups found were Bacillaceae, *Cellulomonas, Microbacterium,* and *Lactobacillus.*

Adeyeba and Akinbo (2002) found a high degree of contamination of solid waste dump sites with bacterial and parasitic agents. The organisms found were *Ascaris lumbricoides, Entamoeba histolytica,* hookworm/strongyle, *Ascaris suum, Ascaris vitulorum, Strongyloides papillosus, Schistosoma suis,* and *Dicrocoelium dendriticum.* Some of these infect humans (see Table 4.1); others infect domestic animals. Some of the bacterial organisms were *Klebsiella* species, *Escherichia coli, Proteus* species, streptococci, and other Gram-positive microorganisms.

Bioaerosols were also detected in airborne dust. They included endotoxins and fungi. Danuta et al. (2004) measured the air quality in offices of municipal landfills. They used a six-stage Anderson sampler and found that the concentration of bacterial aerosols ranged from 1×10^3 to 7.3×10^4 CFU/m^3. Outdoors, the range was similar. The fungal units, which could be more significant from a respiratory aspect, ranged from 2.3×10^2 to 7.3×10^3 indoors and 2.0×10^2 to 1.2×10^4 CFU/m^3. The implication to public and worker health was discussed (Danuta et al. 2004). Bioaerosols are discussed in detail in Chapter 8.

There are many sources of pathogens from solid waste that can affect humans, primarily

- Soil: human wastes, solid wastes, animal wastes
- Air: human wastes, solid waste, animal wastes
- Water for bathing and drinking: human wastes, animal wastes
- Plants: solid waste, domestic and wild animal wastes
- Domestic animals: contaminate soil, directly contaminate humans
- Wild animals: plants, soil

Achudume and Olawale (2007) collected and enumerated pathogens from four different dumping sites (solid waste sources). Their results showed the presence of bacterial species, including *Pseudomonas, Micrococcus, Actinomyces, Neisseria,*

TABLE 4.1
Human Pathogens of Significance Reported in the Cited Literature

Organism	Disease
Acinetobacter baumannii	Gram-negative bacteria; cause pneumonia, urinary tract infections, meningitis
Actinobacillus pleuropneumoniae	Parasites or pathogens in mammals, birds, and reptiles
Actinomyces spp.	Gram-positive bacillus; causes infection in dental procedures and oral abscesses Agents: *A. israelii, A. gerencseriae, A. naeslundii, A. odontolyticus, A. viscosus, A. meyer,* and *Propionibacterium propionicum* Normal flora of mouth, gut, genital tract
Aeromonas spp.	In humans and marine animals *Aeromonas hydrophila, Aeromonas caviae, Aeromonas sobria* cause gastroenteritis
Ascaris lumbricoides	Cause bloody sputum, cough, fever, abdominal pain, infection
Bacillus spp.	Genus of Gram-positive bacilli *B. anthracis* causes anthrax; *B. cereus* causes a foodborne illness
Bacillaceae spp.	A family of bacteria that includes the genera *Bacillus* and *Clostridium*, which are generally rod shaped, Gram negative, and spore producing; endotoxin
Bordetella spp.	Species are *B. bronchiseptica, B. pertussis,* and *B. parapertussis* Cause whooping cough, respiratory infections
Cellulomonas spp.	Cause infective endocarditis and osteomyelitis
Enterobacteria	Gram-negative bacteria that include many of the more familiar pathogens, such as *Salmonella, Escherichia coli, Yersinia pestis, Klebsiella,* and *Shigella*
Entamoeba histolytica	Causes amebiasis, acute colitis, bloody diarrhea, abdominal pain
Escherichia coli var. II	Causes digestive tract infections
Klebsiella pneumonia	Causes infections, including pneumonia, bloodstream infections, wound or surgical site infections, and meningitis
Microbacterium spp.	Genus includes pathogens known to cause serious diseases in mammals, including tuberculosis (*Mycobacterium tuberculosis*) and leprosy (*Mycobacterium leprae*)
Micrococcus	May be involved in other infections, including recurrent bacteremia, septic shock, septic arthritis, endocarditis, meningitis, cavitating pneumonia (in immunosuppressed patients)
Neisseria	Two are pathogens, *N. meningitides* and *N. gonorrhea*; often cause asymptomatic infections
Pasteurella haemolytica	Primarily causes a respiratory infection in sheep and goats
Proteus spp.	*P. vulgaris, P. mirabilis,* and *P. penneri*: opportunistic human pathogens *Proteus* includes pathogens responsible for many human urinary tract infections *P. mirabilis* causes wound and urinary tract infections
Pseudomonas spp.	*Pseudomonas aeruginosa* increasingly recognized as an emerging opportunistic pathogen of clinical relevance

(Continued)

TABLE 4.1 (Continued)
Human Pathogens of Significance Reported in the Cited Literature

Organism	Disease
Salmonella spp.	Acute symptoms: Nausea, vomiting, abdominal cramps, diarrhea, fever, headache
	Chronic consequences
	Salmonellosis is infection of the intestines
Serratia plymuthica	Are not common source of infection; nosocomial septicemia has been reported
Staphylococcus aureus	Found on the skin and respiratory system; causes skin infection (e.g., boils), respiratory disease, food poisoning
Streptococcus faecalis	Can cause endocarditis and bacteremia, urinary tract infections, meningitis, and other infections in humans
Streptococcus pyogenes	Is the cause of many important human diseases, ranging from mild superficial skin infections to life-threatening systemic diseases
	Infections typically begin in the throat or skin
Streptococcus pneumoniae	Major cause of pneumonia

Sources: Major sources: Food and Drug Administration (FDA). 2012. Bad bug book—foodborne pathogenic microorganisms and natural toxins handbook. http://www.fda.gov/food/foodsafety/foodborneillness/foodborneillnessfoodbornepathogensnaturaltoxins/badbugbook/default.htm; Centers for Disease Control and Prevention (CDC). 2012. Diseases and conditions. http://www.cdc.gov/diseasesconditions.

Bacillus, and *Klebsiella*. In later work, Achudume and Olawale (2010) showed that *Klebsiella pneumoniae*, *Pseudomonas aeruginosa*, and other nonfermenting heterotrophic microbes were identified in soil underlying urban waste sites in southwestern Nigeria.

A study by Shantha, Sarayu, and Sandhya (2009) in India evaluated bioaerosols from a municipal dumping ground. Depending on the time of the year, the range of bioaerosols was from nil to 10^6 CFU/m^3. In most cases, the values were higher than 10^6. Bacteria in the air were *Salmonella* spp., *Klebsiella pneumoniae*, *Enterobacteria*, and *Pseudomonas* spp. Any of these can result in human disease. It is interesting that with the more recent emphasis on endotoxins and *Aspergillus* species, these were not enumerated.

Flores-Tena et al. (2007) evaluated the air, soil, and leachate in the San Nicolas landfill in Mexico. The authors indicated that landfills are the most common methods of MSW disposal in major Mexican cities. No information was provided on the design of the landfills and their mode of operation, which have a significant effect on both the leachate and the air. They isolated 39 pathogenic and opportunistic Gram-negative bacteria. Ten were pathogenic, 17 were opportunistic, and 2 were plant pathogens. The means of the total bacterial count in soil, leachate, and air samples were 3.0×10^8 CFU/g, 1.5×10^6 CFU/mL, and 4.4×10^3 CFU/m^3, respectively. The organisms found in leachate were *Actinobacillus pleuropneumoniae*, *Bordetella* spp., *Escherichia coli* var. II, and *Acinetobacter baumannii*. Airborne bacteria represented 19 species. *Pasteurella haemolytica* was isolated in all the air samples. Other species detected were *Serratia plymuthica* and *Aeromonas*. There were many other species found, but in smaller amounts.

One of the most comprehensive studies on occupational and environmental health issues of solid waste management was a report written by Cointreau (2006). The data describe infectious diseases, respiratory infections, and cancer. Some of the diseases were from foodborne pathogens, and others were from vectors.

The data presented clearly indicate the lack of recent information on dumps, especially in developing countries. Most of the data describe findings of primary pathogens. Totally lacking is information on bioaerosols, especially *Aspergillus fumigatus*, molds, and endotoxin. Asthma has been on the increase, even in developed countries. This, to some extent, can be attributed to molds and other bioaerosols.

VECTORS

One of the most important aspects of open landfills, dumps, or MSW left in streets is the issue of vectors. Vectors are difficult to control, particularly in developing countries where slums occur, dumps exist, and there is poor sanitation. Solid waste management is key to vector control. Probably the most well-known attempt at vector control is mosquito eradication. Major diseases through vectors are indicated in Table 4.2.

At the beginning of the twentieth century, vector-borne diseases were a significant global health problem (Gubler 2008), and are still a major health problem in developing countries. Vector-borne disease describes an illness caused by an infectious microbe that is usually transmitted to people. However, vertebrates, including foxes, raccoons, skunks, and dogs, which can all transmit the rabies virus to humans via a bite, can also act as vectors. Arthropods account for over 85% of all known animal species, and they are the most important disease vectors. They can infect animals as well as humans. More than 600 different viruses are transmitted by arthropods. The most common vectors are flies, mosquitoes, ticks, lice, fleas, and bugs. Mosquitoes, which primarily transmit parasitic and viral diseases, are the most important arthropod vectors (Gubler 2008). This is especially true in Africa and Asia, where there has been a major effort to eradicate mosquitoes.

The disease agent (e.g., virus and bacteria) is normally found in a reservoir that can be an animal or a physical environment (e.g., soil). Children are at particular risk from vectors. The diseases transmitted by vectors could be viral (e.g., yellow fever, dengue fever, encephalitis); bacterial, such as *Rickettsia typhi* and *Yersinia* species; and parasitic, either helminths or protozoa. Uncollected waste containing stagnant water breeds mosquitoes. Containers, used tires, and similar items will retain water after a rain and are excellent breeding grounds for mosquitoes. It is estimated that 50 to 100 million cases of dengue fever occur each year.

Ahmed (2011) reported on vectors of pathogens from selected refuse dumps in northern Nigeria. The vectors found were the cockroach (*Periplaneta americana*), dung beetles (*Canthon pilularius*), housefly (*Musca domestica*), black garbage fly (*Ophyra leucostoma*), stable fly (*Stomoxys calcitrans*), latrine fly (*Fannia scalaris*), and other species of the Diptera order. The cockroach is known to feed on fecal

TABLE 4.2
Major Diseases through Vectors

Disease/Disease Agent	Illness	Vector	Reference
Yellow fever	Febrile illness, severe liver disease	Mosquito	CDC 2011
Dengue	Febrile illness, severe musculoskeletal pain	Mosquito; transmitting four serotypes of flavivirus	CDC 2011 *Taber's Cyclopedic Medical Dictionary*
Encephalitis	Flu-like symptoms, headache, seizures, inflammation of the brain	Mosquito	Mayo Clinic (2011)
Yersinia	*Y. pestis* causes plague	Rodents	*Sherris Medical Microbiology* (Ryan and Ray 2004)
Rickettsia typhi	Murine typhus: fever, headache, chills nausea, vomiting	Fleas	CDC 2009

Note: CDC = Centers for Disease Control and Prevention.

matter and transmit such diseases as amoebiasis caused by *Entamoeba histolytica,* giardiasis, and *Toxoplasma gondii.*

The two species of beetles are known to transmit *Toxoplasma gondii* and *Cryptosporidium parvum.* The housefly is known to transmit protozoan parasites and bacteria, such as *Salmonella, Shigella, Campylobacter, Escherichia, Enterococcus, Chlamydia,* and other species that produce diseases (Ahmed 2011). The black garbage flies, although not a direct human vector, contaminated food, causing such infections as polio, typhoid fever, dysentery, and food poisoning. The stable fly is known to harbor a variety of pathogens that cause diseases in humans. Several of the other species found can also cause diseases in humans (Ahmed 2011). The article did not report on direct infections from these vectors on diseases. Essentially, it reported on vectors found in Nigerian dumps that could be responsible for human diseases attributed to uncollected waste.

SUMMARY

Municipal solid waste, often referred to as garbage or household waste to differentiate it from industrial wastes, contains pathogens from numerous sources, such as diapers, spoiled food, contaminated clothing, and other material. These may include body fluids and blood, facial tissues, animal and human wastes, and other sources.

In developed countries, most MSW is either landfilled or incinerated. In the United States, over 50% is landfilled, whereas in Europe the majority is incinerated. Landfills are regulated to reduce or prohibit groundwater contamination, and daily landfill cover reduces the dispersion and dissemination of pathogenic bioaerosols.

Recycling, reuse, source separation, and mechanical separation are increasing but often depend on markets for such items as paper, cardboard, glass, and plastics.

In developing nations and emerging industrial nations with large populations, much of the MSW ends up in uncontrolled dumps. These sites are open, and scavengers roam them to seek reusable and resalable items. Consequently, workers or scavengers are exposed to pathogens, resulting in diseases. Not only are the individuals exposed, but also they bring pathogens into the home and public places on their clothing and shoes. Furthermore, MSW is often left in streets and not removed for days, harboring vectors.

Dumps are major sources of vectors such as rats, mosquitoes, and other arthropods. These are major transmitters of diseases. Mosquitoes proliferate in standing water, often found in containers or used rubber tires. Dengue fever, resulting in millions of cases of fever and mortality, especially in children, occurs annually.

Dumps that are located near bodies of drinking water (e.g., streams) contaminate the water, resulting in gastrointestinal diseases. Open dumps are a source of pathogenic bioaerosols, resulting in respiratory diseases in handlers, workers, and persons separating recyclable or reusable items.

The change in demographics and increased urbanization puts additional cost on disposal and proper treatment of MSW in developing countries. In places like India, large families live in crowded conditions, resulting in transmission of diseases among family members.

REFERENCES

Achudume AC and Olawale JT. 2007. Microbial pathogens of public health significance in waste dumps and common sites. *J Environ Biol* 28: 151–154.

Achudume AC and Olawale JT. 2010. Enumeration and identification of Gram-negative bacteria present in soil underlying urban waste-sites in southwestern Nigeria. *J Environ Biol* 31: 643–648.

Adeyeba OA and Akinbo JA. 2002. Pathogenic intestinal parasites and bacterial agents in solid waste. *East Afr Med J* 79: 604–610.

Ahmed AB. 2011. Insect vectors of pathogens in selected undisposed refuse dumps in Kaduna Town, northern Nigeria. *Sci World J* 6: 21–26.

Althaus H, Sauerwald M, and Schrammeck E. 1983. Waste from hospitals and sanatoria. *Zentralbl Bakteriol Mikrobiol Hyg B* 178: 1–29.

Centers for Disease Control and Prevention (CDC). 2009. Outbreak of *Rickettsia typhi* infection—Austin, Texas, 2008. http://www,cdc.gov/mmwr/preview/mmwrhl/mm5845a4.htm

Centers for Disease Control and Prevention (CDC). 2011. Yellow fever. http://www.cdc.gov/yellowfever/

Centers for Disease Control and Prevention (CDC). 2012. Diseases and conditions. http://www.cdc.gov/diseasesconditions

Cointreau S. 2006. *Occupational and Environmental Health Issues of Solid Waste*. Washington, DC: International Bank for Reconstruction and Development/World Bank.

Collins GH and Kennedy DA. 1992. The microbial hazards of municipal and clinical wastes. *J Appl Bacteriol* 73: 1–6.

Danuta OL, Krzysztof U, Wlazlo A, and Pastuszka JS. 2004. Microbial air quality in offices at municipal landfills. *J Occup Environ Hyg* 1: 62–68.

Engelbrecht RS and Amirhor P. 1975. *Inactivation of Enteric Bacteria and Viruses in Sanitary Landfill Leachate*. Urbana, IL: University of Illinois, Department of Civil Engineering. Federal Communication Commission. Ntisl BB 234.

Flores-Tena FJ, Guerrero-Barrera AL, Avelar-Gonzalez FJ, Ramirez-Lopez EM, and Martinez-Saldana MC. 2007. Pathogenic and opportunistic Gram-negative bacteria in soil, leachate and air in San Nicolas landfill at Aquascalientes, Mexico. *Rev Latinoam Microbiol* 49: 25–30.

Food and Drug Administration (FDA). 2012. Bad bug book—foodborne pathogenic microorganisms and natural toxins handbook. http://www.fda.gov/food/foodsafety/foodborneillness/foodborneillnessfoodbornepathogensnaturaltoxins/badbugbook/default.htm

Gubler DJ. 2008. The global threat of emergent/reemergent vector-borne diseases. http://www.ncbi.nlm.nih.gov/books/NBK52945/

Khatib IA. 2011. Municipal solid waste management in developing countries: future challenges and possible opportunities. In *Integrated Waste Management*, Vol. 2, ed. Kumar S. Rijeke, Croatia: INTECH.

Mayo C. 2011. Encephalitis. http://www,mayoclinic.com/health/encepholitis/DS00226

Pourcher A, Sutra L, Hebe I, Moguedet G, Bollet C, Simoneau P, and Gardan L. 2001. Enumeration and characterization of cellulolytic bacteria from refuse of a landfill. *FEMS Microbiol Ecol* 34: 229–241.

Ryan KJ and Ray CG, eds. 2004. *Sherris Medical Microbiology*. New York: McGraw Hill.

Shantha R, Sarayu K, and Sandhya S. 2009. Molecular identification of air microorganisms from municipal dumping ground. *World Appl Sci J* 7: 689–692.

Taber's Cyclopedic Medical Dictionary, 22nd edition. Philadelphia, PA: FA Davis Co.

5 Pathogens and Diseases of Sewage Sludge, Septage, and Human Fecal Matter

INTRODUCTION

The collection, disposal, and management of human wastes greatly affect the potential for diseases from pathogens discharged with these wastes (Table 5.1). In developed countries, humans generally discharge fecal matter through toilets into sewage systems or septic systems. The object of the sewer system is to separate the solids from liquids and to disinfect these components before their discharge into the environment. A generalized sewage treatment system is shown in Figure 5.1.

Although this is a general process flow, there are several variations, all with the same objectives. The objectives are to produce a clean and disinfected effluent and usable solids, termed *biosolids*. During the beginning of the process, large objects such as rags, rocks, and sand may cause damage to pumps and other aspects of the process. These are removed and sent to a landfill. A few facilities operate an incinerator, and the ash is landfilled. Following these first two steps, the influent is treated in the primary treatment. This is also a physical process designed to further separate the liquid from solids. The solids are termed *primary sludge* and later combined with sludge produced by secondary treatment, which is a biological treatment. The effluent is relatively clean, but since it can contain pathogens, it requires disinfection. Since many compounds are soluble in water, they may be present in the effluent, and if there is a need to discharge the effluent into a pristine body of water, further treatment may be needed, called tertiary treatment. This was originally designed to remove nitrogen and phosphorus. However, we more recently learned that effluent might contain organic compounds that could have an impact on the environment. These compounds, such as surfactants, are more difficult and expensive to remove. The disinfected effluent is sometimes used to irrigate golf courses or other public works.

The solids are usually thickened and often are placed into sealed tanks called digesters. During digestion, which is an anaerobic process, methane or biogas is produced. This may be burned or reused to produce power for use in the wastewater facility. *Biosolids* is a term given to processed sewage sludge. The main reason for the use of the term is that the term *sludge* is often used without prefacing it with the word *sewage*, so it can be misleading. For example, the medical profession uses sludge for blockage of the intestinal system. The term *biosolids* is specific. The

TABLE 5.1
Some Bacteria Found in Wastewater, Sludge, and Biosolids and On-Site Soil Transmission Systems and the Diseases They Transmit

Bacteria	Disease
Salmonella spp. (approximately 1,700 types)	Salmonellosis
	Gastroenteritis
Salmonella typhi	Typhoid fever
Mycobacterium spp.	Tuberculosis, leprosy
Shigella spp.	Dysentery
Yersinia spp.	Gastroenteritis
Vibrio cholerae	Cholera
Campylobacter jejuni	Gastroenteritis
Escherichia coli (pathogenic strains)	Gastroenteritis
Helicobacter pylori	Duodenal and gastric ulcers
Yersinia enterocolitica	Diarrhea (watery or bloody), fever
Aeromonas hydrophila	Gastroenteritis, cellulitis, myonecrosis, eczema
Leptospira	Kidney damage, meningitis, liver failure, respiratory stress, death
Brucella	Brucellosis, fever, headache, fatigue, anorexia
Viruses	
Enteroviruses	
Hepatitis A	Infectious hepatitis, liver disease
Hepatitis E	Liver inflammation
Reoviruses	Respiratory and gastrointestinal infections
Rotaviruses	Diarrhea, vomiting, abdominal pain, fever
Adenoviruses	Respiratory infections
Caliciviruses	
Norwalk	
Snow	

biosolids, produced after anaerobic digestion, are usually dewatered. These biosolids are reused as a fertilizer in land application. Land application immediately after digestion may be used directly through either spraying or spreading. Prior to land application, the biosolids must be stabilized to avoid odors and further reduce pathogens. Stabilization can be achieved by composting, alkaline treatment, or heat drying. Biosolids management is regulated by the federal government in a set of regulations termed 40 CFR 503 (Epstein 1997, 2003; National Research Council [NRC] 2002).

A septic system (on-site disposal system, OSDS) is used where a wastewater collection system is not available. In communities where a sewage treatment plant and a collection system are not available, individual homes or several homes combined would have a septic system. The septic system would consist of a tank that receives household wastes from sinks, bathtubs, and toilets. The tank is designed to allow

Pathogens and Diseases of Sewage Sludge, Septage, and Human Fecal Matter 43

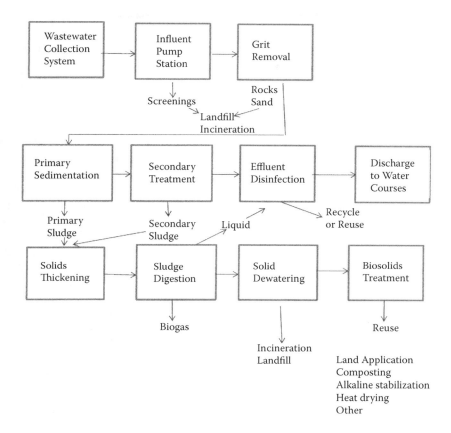

FIGURE 5.1 Generalized sewage treatment system.

solids to settle to the bottom. These solids need to be pumped out periodically. When the tank is full, the liquid portion is passed through perforated pipes into the soil. Fortunately, if the system is designed properly and the soil conditions are right, the pathogens will not survive or reach groundwater. In the United States, approximately 20% of the population in 1998 used septic systems (Meschke and Sobsey 1999). Contamination of groundwater can occur when a system fails and is implicated in up to 40% of groundwater-attributed outbreaks (Cogger 1988).

In developing countries, collection and treatment systems or septic systems are not available or used. Consequently, fecal matter is often discharged outdoors, not even into a pit or outhouse. Children or adults are exposed to these pathogens, and diseases occur. Fecal-to-oral contamination is common. Disease prevention is difficult. Furthermore, as was pointed out, vectors such as flies and mosquitoes can transmit diseases from this source.

Diarrheal stool specimens have been used to identify microbial pathogens. Therefore, the discharge of fecal matter from children with diarrhea can result in infections.

PATHOGENS AND DISEASES

Wastewater contains pathogens from human and animal wastes discharged into the sewer system. In addition, surface runoff will contain animal and avian pathogens. Human fecal matter discharged on soil surfaces, as commonly occurs in developing countries, can directly contaminate individuals or drinking water supplies. Pathogens in runoff can enter bathing waters as well and infect humans.

Global and regional conditions can also affect the type and numbers of certain pathogens. The mobility of our society, ease of travel, and influx of individuals from developing countries, especially from semitropical and tropical regions, increase the likelihood of both numbers and types of parasites entering wastewater. Increased numbers of pathogens with disease resistance as a result of mutation and the improper use of antibiotics are becoming evident. Infections with *Escherichia coli* O157H7 and *Helicobacter pylori* and mad cow disease (bovine spongiform encephalopathy, BSE) are more prevalent, and not much is known about their reduction or elimination during wastewater treatment. The survival of human immunodeficiency virus (HIV) and other pathogens in wastewater has been studied and reported (Casson 1992; Chauret, Springthorpe, and Sattar 1999; Epstein 2003).

Food contamination occurs more frequently. Each year, 48 million people in the United States become sick from contaminated food. Common contaminating culprits include bacteria, parasites, and viruses. Symptoms range from mild to serious, including

- Upset stomach
- Abdominal cramps
- Nausea and vomiting
- Diarrhea
- Fever
- Dehydration

Harmful bacteria are the most common cause of foodborne illness. Foods may have some bacteria on them when you buy them. Raw meat may become contaminated during slaughter. Fruits and vegetables may become contaminated when they grow or are processed (MedlinePlus, nih.gov).

Each year, more than 9 million foodborne illnesses are estimated to be caused by major pathogens acquired in the United States. Using data from outbreak-associated illnesses for 1998–2008, we estimated annual US foodborne illnesses, hospitalizations, and deaths attributable to each of 17 food commodities. We attributed 46% of illnesses to produce and found that more deaths were attributed to poultry than to any other commodity (Painter et al. 2013).

Pathogens can be grouped into two major categories. Primary pathogens can invade and infect healthy humans. Secondary pathogens invade and infect debilitated or immunosuppressed individuals. The important human enteric microbial pathogens in on-site soil transmission systems are shown in Table 5.2 (Meschke and Sobsey 1999).

TABLE 5.2
Important Human Enteric Microbial Pathogens

- Bacteria
    - *Salmonella* spp.
    - *Campylobacter* spp.
    - *Escherichia coli*
    - *Helicobacter pylori*
    - *Aeromonas hydrophila*
    - *Yersinia enterocolitica*
    - *Vibrio cholerae*
    - *Brucella leptospira*
    - *Mycobacteria* spp.
    - *Shigella* spp.
- Viruses
    - Enteroviruses
        - Polio
        - Echos
    - Hepatitis A virus
    - Hepatitis V virus
    - Reoviruses
    - Rotaviruses
    - Adenoviruses
    - Caliciviruses
        - Norwalk
        - Snow
- Protozoan
    - *Cryptosporidium parvum*
    - *Cyclospora cayetanensis*
    - *Giardia lamblia*
    - *Entamoeba histolytica*
    - *Balantidium coli*
    - *Microsporidia*
    - *Toxoplasma gondii*
- Helminth
    - *Ancylostoma duodenale*
    - *Ascaris lumbricoides*
    - *Enterobius vermicularis*
    - *Necator americanus*
    - *Strongyloides stercoralis*
    - *Trichuris trichiura*
- Enteric viruses
- Protozoa
- Helminths
    - Nematodes (roundworms)
    - Cestodes (tapeworms)

Source: Modified from Meschke JS and Sobsey MD. 1999. Microbial pathogens and on-site soil treatment systems. http://.ces.bcsu.edu/plymouth/septic/98meschke.html

Examples of secondary pathogens are

- *Klebsiella* sp.
- *Listeria*
- Molds
- Fungi
    - *Aspergillus fumigatus*
    - *Penicillium*
- Actinomycetes
- Endotoxin

REFERENCES

Casson LW. 1992. HIV survivability in wastewater. *Water Environ Res* 64: 213.
Chauret C, Springthorpe S, and Sattar S. 1999. Fate of *Cryptosporidium* oocysts, *Giardia* cysts and microbial indicators during wastewater treatment and anaerobic sludge digestion. *Can J Microbiol* 45(3): 257–262.
Cogger, C. 1988. On-site septic systems: the risk of groundwater contamination. *J Environ Health* 51(1), 12–16.
Epstein E. 1997. *The Science of Composting*. Lancaster, PA: Technomic.
Epstein E. 2003. *Land Application of Sewage Sludge and Biosolids*. Boca Raton, FL: CRC Press.
MedlinePlus. nih.gov Foodborne illnesses, foodborne diseases. Atlanta, GA: Centers for Disease Control and Prevention.
Meschke JS and Sobsey MD. 1999. Microbial pathogens and on-site soil treatment systems. http://.ces.bcsu.edu/plymouth/septic/98meschke.html.
National Research Council (NRC). 2002. *Biosolids Applied to Land*. Washington, DC: National Research Council of the National Academies Press.
Painter JA, Hoekstra RM, Ayers T, Tauxe RV, Braden CR, Angulo FJ, and Griffin PM. 2013. Attribution of foodborne illnesses, hospitalizations, and deaths to food commodities by using outbreak data, United States. 1998–2008. *Emerg Infect Dis* 19(3): 407–415.

6 Pathogens in Soils

INTRODUCTION

The importance of the role of pathogens in soil depends greatly on the method of collection and treatment of human waste materials. The potential health impacts to populations greatly depends on the availability of infrastructure to remove and treat human wastes, potential for contact with human wastes, control and prohibition of vectors to transmit diseases, and education related to the avoidance of disease potential.

The early literature regarding the persistence and movement of pathogens dealt with land application of sewage, effluent, or low-solids sewage sludge (Sepp 1971). This literature indicated that the use of raw sewage could result in animal or human health infections (Bicknell 1972; Dunlop, Twedt, and Wang 1951; Dunlop and Wang 1961). Currently, the practice of applying untreated sewage or sewage sludge is prohibited in the United States, Canada, and Europe. The United States Environmental Protection Agency (USEPA) 503 regulations allow land application of biosolids (treated sewage sludge) as either class A or class B material. Class A treatments are intended to significantly reduce pathogens to virtually nondetected limits. Class B treatments allow land application with crop restrictions and a regulated concentration of fecal coliforms (see Epstein 1997). With all the regulations and efforts by agencies such as the US federal government, there are still numerous incidences of foodborne and waterborne diseases. From 1990 to 2003, produce caused the most foodborne disease in the United States. In England and Wales, 6.4% and 10.1% of all disease outbreaks with a known food vehicle occurred from 1993 to 1998 and 1999 to 2000, respectively. Some of the foodborne diseases could be the result of soil contamination (Brandl 2006).

The survival of pathogens in soils depends on the pathogen surviving the wastewater treatment processes and biosolids treatment, method of land application, defecation on soil surfaces, soil conditions, and environmental conditions. In large communities, many of the biosolids processes can result in effective disinfection, and the application of these biosolids does not represent a health hazard. Public perception may be significantly negative so that applying biosolids to the land for certain food chain crops must be avoided. Figure 6.1 shows the potential routes of pathogen transmission to humans and animals. Food chain crops that are cooked or processed would have no potential for infection. These would include crops for oil (soybeans, sunflower, and canola) or canned foods. Food crops that would not come in contact with biosolids, such as fruit trees (citrus, nut, pears, apples, etc.), would also not harbor pathogens and would not be a potential source of infection. Non-food

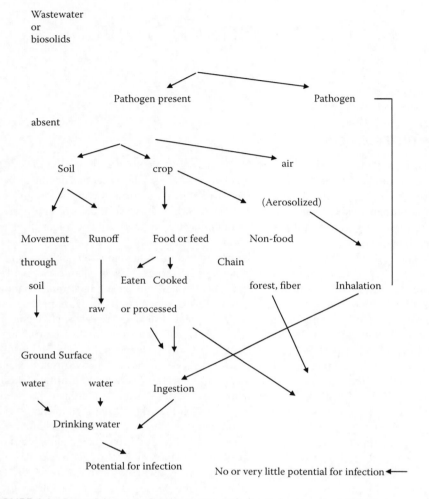

FIGURE 6.1 Potential routes of pathogen transmission to humans and animals.

chain crops such as fiber (cotton) or forest would also not present a hazard to humans or animals.

Although class B biosolids can contain pathogens, the USEPA provided site restrictions to preclude potential impacts to human and animal health and environmental consequences. These site restrictions were based on the potential survival of pathogens in the environment.

When low-solid biosolids are applied by spraying on land, there is the potential for pathogens to be aerosolized. Workers in particular may be subjected to these pathogens and become infected. Contamination of water resources, especially drinking water, can result from the movement of pathogens through the soil or in runoff into surface waters.

Since pathogens survive the wastewater treatment processes (primary and secondary treatment), land application of sewage sludge directly from these processes needs to be avoided or restricted to land management systems that would have minimal potential for environmental contamination or health hazards to humans and animals. Further treatment, such as digestion, composting, alkaline stabilization, or heat drying, increases the opportunities for land application. Composted products can be used as soil-conditioning materials for agricultural and horticultural applications, such as landscaping, nurseries, parks, public work projects, and home gardens; alkaline-stabilized biosolids can be used as lime products and soil amendments for agriculture and horticultural practices, such as turf production and certain public works projects; and heat-dried materials are often used as substitutes for chemical fertilizers.

Pathogen survival in soils depends on several factors:

- Climatic and microclimate effects
- Soil physical properties
- Soil chemical properties
- Soil biological properties, especially antagonistic microbial populations

The survival of pathogens in soils and their potential movement through soils to groundwater depends on several edaphic and climatic factors. The most important are soil moisture, pH, temperature, organic matter, soil texture, soil permeability, sunlight, and antagonistic microflora.

Exposure to sunlight and ultraviolet light will destroy pathogens on soil surfaces. Desiccation and temperature will destroy organisms on the soil surface. Physical properties of soil such as soil moisture, soil temperature, organic matter, and ionic strength can affect pathogen survival and movement through soils.

Rudolfs, Falk, and Ragotzkie (1950) summarized the literature prior to 1950 on the occurrence and survival of enteric, pathogenic, and relative organisms in soil, water, sewage, sludges, and vegetation. They concluded from the early literature survey that the following soil factors affect the survival of pathogenic microorganisms:

- Type of organism: *Escherichia coli, Eberthella typhosa,* and *Mycobacterium tuberculosis* appear to be most resistant.
- Soil moisture content: Longevity is greater in moist soils. High soil moisture-holding capacity increases longevity.
- Soil pH: Neutral soils favor longevity.
- Organic matter: The type and amount of organic matter may serve as food and energy sources to sustain or allow bacteria to increase.
- The presence of other microorganisms reduces the presence or concentration of pathogenic organisms.
- Temperature: There is longer survival at low temperatures.

Other literature reviews were provided by Dunlop (1968) and Sepp (1971). Gerba, Wallis, and Melnick (1975) reviewed the literature prior to 1975 and substantiated the

TABLE 6.1
Survival of Bacteria in Soils

Organism	Soil	Temperature (°C)	Survival Time (Days)
Salmonella spp.	Soil	—	15–7280
Salmonella serotype Typhimurium	Soil	Summer sun	<28
Salmonella serotype Typhimurium	Soil	Summer shade	<70
Salmonella typhi	Soil	—	1–120
Salmonella typhi	Sandy soil	16–17	<29 to <58
Salmonella typhi	Soil moist	—	<80
Salmonella typhi	Soil dry	—	<24
Tubercle bacilli	Soil	—	>189
Leptospira	Soil	—	15–43
Coliform	Soil surface	—	38
Streptococcus spp.	Soil	—	35–63
Fecal streptococci	Soil	—	26–77

Source: Modified from Rudolfs W, Falk LL, and Ragotzkie RA. 1950. Literature review of the occurrence and survival of enteric pathogenic, and relative organisms in soil, water, sewage, and sludge and on vegetation. *Sewage Indust Wastes* 22: 1261–1281.

TABLE 6.2
Survival Time for Some Bacteria in Soil

Bacteria	Soil	Moisture	Temperature (°C)	Survival
Streptococci	Loam	?	?	9–11 weeks
Streptococci	Sandy loam	?	?	5–6 weeks
Salmonella typhi	Various soils	?	22	2–400 days
Bovine tubercle bacilli	Soil and dung	?	?	Less than 178 days
Leptospires	Varied	Varied	Summer	12 hours to 15 days

Source: Modified from Feachem JB, Bradley DJ, Garelick H, and Mara DD. 1980. *Appropriate Technology for Water Supply and Sanitation.* Washington, DC: World Bank.

soil factors of Rudolfs, Falk, and Ragotzkie (1950). They indicated that sunlight also reduced survival time on soil surfaces. The early data on the survival of pathogens in soil are shown in Table 6.1 as provided by Rudolfs, Falk, and Ragotzkie (1950). Tables 6.1 and 6.2 provide early data on the survival of some pathogens in soils.

One of the most comprehensive early reviews on the persistence of pathogens in soil was conducted by Feachem et al. (1980). Some of his data are shown in Tables 6.3 to 6.6, which provide information on pathogen persistence in soils. Table 6.3 shows

TABLE 6.3
Survival of Enteroviruses in Soil

Soil Type	pH	Moisture (%)	Temperature (°C)	Days of Survival (Less than)
Sterile, sandy	7.5	10–20	3–10	130–170
		10–20	18–23	90–110
	5.0	10–20	3–10	110–150
		10–20	18–23	40–90
Nonsterile, sandy	7.5	10–20	3–10	110–170
		10–20	18–23	40–110
	5.0	0–20	3–10	90–150
		10–20	18–23	25–60
Sterile, loamy	7.5	10–20	3–10	70–150
		10–20	18–23	70–110
	5.0	10–20	3–10	90–150
		10–20	18–23	25–60
Nonsterile, loamy	7.5	10–20	3–10	110–150
		10–20	18–23	70–110
	5.0	10–20	3–10	90–130
		10–20	18–23	25–60
Nonsterile, sandy	7.5	10–20	18–23	15–25

Source: Data from Feachem JB, Bradley DJ, Garelick H, and Mara DD. 1980. *Appropriate Technology for Water Supply and Sanitation.* Washington, DC: World Bank.

the survival of enteroviruses, the majority of which do not survive more than 150 days in soil. (Feachem et al. 1980). Table 6.4 shows that although parasites in soil rarely survive for a few weeks, hookworms may survive up to six months.

Reddy, Khaleel, and Overcash (1981) provided a comprehensive review of the behavior and transport of microbial pathogens and indicator organisms in soils treated with organic wastes. They also concluded that the most important factors affecting the die-off rate were temperature, moisture, pH, and method of waste application. Die-off doubled for temperature increases of 10°C and increased for decreases in soil moisture. Retention of microorganisms increased with an increase in clay content of the soil. Table 6.6 shows the average die-off rate constants for selected indicator organisms and pathogens.

The decay die-off rates for *Salmonella* serotype Typhimurium and fecal coliform were in the range that Reddy et al. (1981) had shown. The *Shigella sonnei* decay coefficient rates were lower. Evans et al. (1995), with column studies, found that *S. sonnei* and fecal coliforms survived longer on average than *Salmonella* serotype Typhimurium. Casson (1996) also conducted a laboratory study and showed that two log reductions were found after 10 to 20 days. Decay rates ranged between 0.08 log per day and 0.4 log per day.

TABLE 6.4
Survival Time for Some Parasitic Worms in Soil

Worm	Soil	Moisture	Temperature (°C)	Survival
Hookworm larvae	Sand	?	Room temperature	Less than 4 months
	Soil	?	Open shade, Sumatra	Less than 6 months
	Soil	Moist	Dense shade	9–11 weeks
			Moderate shade	6–7.5 weeks
			Sunlight	5–10 days
	Soil	Water covered	Varied	10–43 days
	Soil	Moist	0	Less than 1 week
			16	14–17.5 weeks
			27	9–11 weeks
			35	Less than 3 weeks
			40	Less than 1 week
Hookworm ova (eggs)	Heated soil with night soil	Water covered	15–27	9% survival after 2 weeks
	Unheated soil with night soil	Water covered	15–27	3% survival after 2 weeks
Roundworm ova	Sandy, shade		25–36	31% dead after 54 days
	Sandy, sun		24–38	99% dead after 15 days
	Loam, shade		25–36	3.5% dead after 21 days
	Loam, sun		24–38	4% dead after 21 days
	Clay, shade		25–36	2% dead after 21 days
	Clay, sun		24–38	12% dead after 21 days
	Humus, shade		25–36	1.5% dead after 22 days
	Clay, shade		22–35	More than 90 days
	Sandy, shade		22–35	Less than 90 days

Source: Modified from Feachem JB, Bradley DJ, Garelick H, and Mara DD. 1980. *Appropriate Technology for Water Supply and Sanitation*. Washington, DC: World Bank.
Note: Data from other scientists are shown in Table 6.5.

TABLE 6.5
Survival of Some Viruses in Soil

Organism	Soil	Application System	Temperature (°C)	Survival Time (Days)	Reference
Poliovirus	Sand dunes (dry)				
Poliovirus type 1		Effluent		31	
			33	>40 days	Palfi 1972
			30	Reduced by 4 logs in 30 days	Moore, Sagic, and Sorber 1978
Poliovirus 1		Secondary effluent	Winter Summer	89 days 11 days	
		Activated sewage sludge	Winter Summer	96 days 7 days	Tierney, Sullivan, and Larkin 1977
Coxsackievirus	Sandy soil and clay	Dewatered Anaerobically digested	Winter	23 weeks	

TABLE 6.6
Die-off Rate Constants (Day^{-1}) for Selected Indicator Organisms and Pathogens in a Soil-Water-Plant System

Organism	Average	Maximum	Minimum	Standard Deviation ±	Coefficient of Variation (%)	No. of Observations
Escherichia coli	0.92	6.39	0.15	0.64	179	26
Fecal coliforms	1.53	9.10	0.07	4.35	283	46
Fecal streptococci	0.37	3.87	0.05	0.69	188	34
Salmonella spp.	1.33	6.93	0.21	1.70	128	16
Shigella spp.	0.68	0.62	0.74	0.06	9	3
Staphylococcus spp.	0.16	0.17	0.14	0.02	14	2
Viruses	1.45	3.69	0.04	1.44	99	11

Source: After Reddy KR, Khaleel R, and Overcash MR. 1981. Behavior and transport of microbial pathogens and indicator organisms in soils treated with organic wastes. *J Environ Qual* 10(3): 255–266.

BACTERIA

Shortly after the discovery of *Eberthella typhosa* as the causative organism for typhoid, attempts were made to determine its survival in soils (Rudolfs, Falk, and Ragotzkie 1950). Melick (1917) reported that *E. typhosa* survived from 29 to 58 days depending on the soil type. In a sandy soil, survival lasted for 74 days. Kliger (1921) indicated that moist alkaline conditions in soils were most favorable for the survival of pathogens. Beard (1938, 1940) reported that soil water-holding capacity, temperature, precipitation, sunlight, soil pH, and soil organic matter all affected the survival of typhoid bacillus. The data showed that the survival of *Salmonella typhosa* was greatest in soils during the rainy season. In sand, where drying is more rapid, the organism survived for a short time, between 4 and 7 days; however, in soils that retained moisture, the organism persisted for longer than 42 days.

Van Dorsal, Geldreich, and Clarke (1967) found that there was a greater die-off rate for *Escherichia coli* and *Streptococcus faecalis* in soil plots exposed to the sun than for those in the shade. The authors also reported that 90% of fecal coliforms in the soil were reduced in 3.3 days in the summer and 13.4 days in the winter. Bacteria can move through soils to great depths. Romero (1970) reported that after 2 days, fecal coliforms and fecal streptococci organisms were observed to travel over 500 m (1,500 ft) after the application of tertiary treated wastewater. This movement occurred in coarse gravel. Several other early authors reported that bacteria could move through soils to depths ranging from less than 1 to 830 m. The lower soil depths were in sand, sandy gravel, and gravel. The majority of the studies have shown that the movement of bacteria in soils is restricted to less than 30 ft and should not percolate into groundwater (Butler, Orlob, and McGauhey 1954; McGauhey and Krone 1967). These studies have been primarily through the use of wastewater or biosolids at low concentrations. Pathogens in dewatered or high-biosolid material applied to land will not likely leach out and move through the surface soil to groundwater. The application of dewatered biosolids on the movement of bacteria is markedly different from that for wastewater. Surface application, including tilling or incorporating biosolids into the upper 15 cm (6 in.), greatly reduces the survival of bacteria and movement. Andrews, Mawer, and Matthews (1983) found that when biosolids were injected into the soil, 90% were inactivated after 17 days in the winter and 3.7 days in the summer.

Sorber and Moore (1986) reviewed the literature prior to 1986 and concluded that quantitative data describing pathogen survival or transport in biosolids-amended soil were extremely limited. Their data are shown in Table 6.7.

Generally, some salmonellae bacteria and indicator organisms survived for several weeks. Median die-off rates for indicator bacteria, fecal coliforms, fecal streptococci, and total coliforms were less than those observed for *Salmonella*. The literature review presented by Sorber and Moore (1986) showed that, with one exception, there was a 90% reduction in *Salmonella* within 3 weeks of biosolid application. These studies were conducted with both indigenous and seeded organisms (Andrews, Mawer, and Matthews 1983; Jones et al. 1983; Kenner, Dotson, and Smith 1971; Larkin et al. 1978). These authors indicated that seeded *Salmonella* organisms often showed higher persistence. They suggested that could be the result of very high

TABLE 6.7
Survival of Several Microorganisms in Soil

Organism	Depth (cm)	Die-off Rate: T_{90}[a] Days				Die-off rate: T_{99}[b] Days			
		Minimum	Maximum	Median	Observations[c]	Minimum	Maximum	Median	Observations
Salmonella	0–5	6	61	12	11	11	45	22	8
	5–15	4	22	15	8	7	45	30	6
Fecal streptococci	0–5	7	28	17	10	14	63	24	8
	5–15	NA[d]	NA	NA	NA	NA	NA	NA	NA
Fecal coliforms	0–5	7	84	25	19	12	165	60	16
	5–15	4	49	16	10	9	56	32	9
Total coliforms	0–5	16	170	40	7	28	350	155	4
	5–15	35	70	42	3	NA	NA	NA	NA
Viruses	0–5	<1	30	3	9	2	52	6	6
	5–15	30	56	30	3	60	100	60	3
Parasites	0–5	17	270	77	11	68	500	81	5
	5–15	NA	NA	NA	NA	NA	NA	NA	NA

Source: From Sorber CA, and Moore BE. 1986. Survival and transport of pathogens in sludge-amended soil. In *Proceedings of the National Conference on Municipal Treatment Plant Sludge Management.* Orlando, FL: Information Transfer, 25–32.

[a] T_{90} = 90% reduction within the days indicated.
[b] T_{99} = 99% reduction within the days indicated.
[c] Obser., observations.
[d] NA, data not available.

concentrations, 10^6 to 10^9 per liter, that were land applied. In field studies, indigenous *Salmonella* generally persisted for less than 2 months, with few positive recoveries reported for as long as 3 to 5 months. Strauch, Konig, and Evers (1981) evaluated the survival of seeded salmonellae in biosolids applied to forestland. The soils were sand and marl. The average temperature was 8.3°C, and average precipitation was 739 mm. Salmonellae survived from 270 to 640 days. Watson (1980), in a study in England, found that organic matter, pH, temperature, and the physical state of the organism affected salmonellae survival when digested biosolids were applied to land. Salmonellae concentrations dropped from 100 million to zero in 42 to 49 days.

The sieving effect of soil, which is affected by particle size, texture (i.e., clay vs. sand), and adsorption, can greatly reduce bacterial movement. Alexander et al. (1991) studied the factors affecting the movement of bacteria through soil. They measured sorption partition coefficient, hydrophobicity, net surface electrostatic charge, zeta potential, cell size, encapsulation, and flagellation of the cells using 19 different bacterial strains. The results indicated that adsorption greatly contributed to the retention of bacteria and that bacterial movement through aquifer sand was enhanced by reducing the ionic strength of the in-flowing solution. Cell density and flow velocity also affected bacterial movement. The data indicated that the potential for bacterial contamination of groundwater from the application of biosolids was minimal. Studies inoculating bacteria in sterile and nonsterile soil showed that pathogens were suppressed by the presence of other soil organisms. Bryanskaya (1966) showed that actinomycetes suppressed the growth of salmonellae and dysentery bacilli.

Pepper et al. (1993) conducted both laboratory and field studies using total coliforms, fecal coliforms, and fecal streptococci organisms. They found that soil moisture, soil temperature, and soil texture affected the survival of these indicator organisms. Survival of organisms increased with increasing soil moisture and clay content and decreased with increasing soil temperature. Under field conditions when soil moisture increased as a result of rainfall, regrowth of indicator organisms occurred.

Although concentrations of fecal coliforms, fecal streptococci, and salmonellae decreased through an extended hot, dry summer period and were not detected, repopulation occurred after precipitation (Gibbs et al., 1997). The authors indicated that despite apparent die-off of salmonellae, land to which biosolids have been applied may be subject to salmonella repopulation. Management needs to take this into account to protect public health.

Land application of biosolids can result in runoff and potential contamination of surface waters. This would be especially true if the biosolids were not incorporated into the soil prior to rainfall (Evans et al. 1995). Land application of biosolids to highly porous soils following significant amounts of precipitation could result in some movement of pathogenic organisms for several meters. However, unless the groundwater levels are very shallow, there is little potential for contamination of groundwater. Biosolid application modifies soil properties, which increases the retention and removal of pathogens. Increased organic matter will decrease water percolation and increase water retention in sandy or gravelly soils. Biosolid application modifies pH, which could affect bacterial survival. This is especially true if the pH is increased through liming. The increased organic matter from biosolid application enhances the indigenous microbial population, which could result in pathogen inactivation.

VIRUSES

Data on virus movement in soil from the application of biosolids are meager. However, considerable information is available as a result of effluent application to land. Application of viruses in effluents offers much greater potential for movement through soils and therefore represents much worse scenarios. Movement with a liquid medium is more rapid than by leaching from a solid matrix. Furthermore, viruses are adsorbed on the solid surfaces and less apt to leach. The organic matter in biosolids would also affect the adsorption of viruses. Survival and movement of viruses through soil are greatly affected by soil properties. Viruses generally do not survive long outside their hosts. They contain a nucleic acid core surrounded by proteins. Viruses are electrically charged colloidal particles and thus are capable of adsorbing onto soil surfaces. Many studies utilized bacteriophages as models. A bacteriophage is a virus with specific affinity for bacteria (Stedman's 1977).

The early studies of adsorption of viruses onto soil surfaces were reviewed by Bitton (1975). Drewry and Eliassen (1968) studied virus retention in soils and concluded that virus adsorption was affected by the soil–water system pH. Carlson et al. (1968) studied the adsorption of bacteriophage T_2 and type 1 poliovirus to kaolinite, montmorillonite, and illite clays. The type and concentration of cations present in soil water affected sorption of viruses under similar ionic conditions. Kaolinite and montmorillonite adsorbed the same amount of viruses. Illite required twice as much salt to attain the same binding capacity. These authors concluded that the surface exchange capacity determined by the surface density and clay particle geometry was important in the adsorption process. Adsorption is more rapid at a lower pH.

Bagdasaryan (1964) reported that enteroviruses survived in loamy and sandy loam soils for prolonged times. The adsorption of viruses to soil may prolong their survival (Gerba, Wallis, and Melnick 1975). Tierney, Sullivan, and Larkin (1977) also found that poliovirus 1 viruses inoculated in raw and activated sludge survived in soils for up to 96 days in the winter. Damgaard-Larsen et al. (1977) found that it took 23 weeks during a normal Danish winter to inactivate 10^6 $TCID_{50}$/g of coxsackievirus B3. Under warm, humid Florida conditions, Farrah, Scheuerman, and Bitton (1981) reported a 2 $\log_{10}$ drop in titer of indigenous viruses in biosolids-amended soil. Gerba (1983) indicated that virus inactivation occurred in the top few centimeters of soil where drying and radiation forces were greatest. Bitton, Pancorbo, and Farrah (1984) evaluated virus transport and survival after land application of biosolids. Viral strains of poliovirus type 1 and echovirus type 1 were mixed with anaerobically and aerobically digested biosolids, applied to the soil, and mixed with the top 2.5 cm of soil. Neither the poliovirus nor echovirus was detectable in soil after being exposed for 8 days to dry fall weather conditions. Under summer weather conditions in Florida, poliovirus was detectable in the soil for 35 days.

Generally, viruses are adsorbed on clays, and the adsorption capacity increased with clay content, cation exchange capacity, specific surface, and organic matter. It has been shown that virus adsorption in natural soils follows the Freundlich isotherm:

$$q = K_F C^{1/n}$$

where q is the amount of virus (PFU/g of soil); C is the concentration of the virus at equilibrium (in PFU/mL solution); K_F is the Freundlich constant determined by the y-axis intercept from a plot of q versus log C; and $1/n$ is the slope of the line as determined by the plot.

Burge and Enkiri (1978) showed that the amount of virus adsorbed by five soils was linearly related to the square root of time. There was a high negative correlation with pH, as would be expected due to their amphoteric nature. The lower the soil pH was, the more positively charged the virus particles were.

Viruses are removed from percolating water by adsorption on soil particles. Lance (1977) found that poliovirus type 1 essentially remained in the top 5 cm. Other factors are soil type, iron oxides, pH, cations, and virus type. Gilbert, Gerba, et al. (1976) reported that human bacterial and viral pathogens are largely removed as sewage effluent percolates through the soil. The viruses measured included polio, echo 15, coxsackie B4, reovirus 1 and 2, and undetermined types. The bacterial indicators and bacterial pathogens were fecal coliforms, fecal streptococci, and *Salmonella* spp. Gilbert, Rice, et al. (1976) also reported that human viral pathogens do not move through soil into groundwater. Lue-Hing, Sedita, and Rao (1979) did not find any evidence of viral contamination of runoff, surface water, or groundwater at the Fulton County, Georgia, biosolids application site. Although the adsorption of viruses to clays precludes their movement to groundwater, Shaub et al. (1975) have shown that adsorbed viruses were still infectious.

Straub, Pepper, and Gerba (1992) measured the inactivation rate ($k = \log^{-10}$ reduction per day) of poliovirus type 1 and bacteriophages MS2 and PRD-1, in a laboratory study using two desert soils. Biosolids were added to a Brazito sandy loam and Pima clay loam. They found that, under constant moisture and temperature, temperature and soil texture were the most important factors controlling inactivation. As the temperature increased from 15°C to 40°C, the inactivation rate for poliovirus and bacteriophage MS2 increased, whereas for the bacteriophage PRD-1 a significant increase in inactivation occurred only at 40°C. Clay soils afforded more protection to all three viruses than sandy soils. Reduction in moisture content to less than 5% completely inactivated all three viruses within 7 days at 15°C. Thus, a combination of moisture reduction and high temperature is effective in virus inactivation. These studies under laboratory conditions using soil columns and constant parameters can provide indication of possible trends but should not be taken as definitive behavior of organisms in the environment. Soils undergo fluctuations in moisture and temperature. These fluctuations, especially desiccation and high surface temperatures, will destroy pathogens.

PARASITES

Sorber and Moore (1986) reported that parasites persisted the longest of most organisms in soils. Gerba (1983) indicated that air, desiccation, and sunlight will result in rapid die-off. Protozoan cysts are highly sensitive to drying and are expected to survive for only a few days in most soils. *Entamoeba histolytica* has been reported to survive 18–24 hours in dry soil and 42–72 hours in moist soil (Kowal 1982). Helminth eggs and larvae can survive in soil for longer periods. Under favorable conditions of moisture, temperature, and sunlight, *Ascaris*, *Trichuris*, and *Toxocara* can remain viable and infective for several years (Little 1980). *Ascaris* eggs can survive up to 7 years in soil (Sepp 1971). However, Strauch, Konig, and Evers (1981) reported that *Ascaris* eggs are dependent on a host for survival and thus died fairly quickly in both winter and summer when seeded biosolids were applied to forestland. In the East Bay Municipal Utility District (EBMUD, Oakland, California) study, 12.9% of the soil samples contained viable helminth ova 3 years after biosolid application (Theis, Bolton, and Storm 1978). Feachem et al. (1950) reported that hookworms can survive for up to 6 months. Babaeva (1966) indicated that *Taenia* could survive from several days to 7 months.

SUMMARY

Although the data on pathogen survival in soils are highly variable, it is evident that most pathogens do not survive in soils for a great length of time. The soil environment is generally hostile to pathogens. Desiccation, soil temperature, and pH affect their survival. Decreases in soil moisture have resulted in a greater pathogen die-off rate. Increases in soil temperature increase the die-off rate of pathogens. Soil acidity also lowers the rate of pathogen survival. Organic matter and clay affect retention, especially of viruses. Generally, most pathogens are retained in the upper 5 to 15 cm of the soil. The data show that bacteria and viruses do not survive in soils for extensive periods. Thus, the potential for bacteria and viruses to move through soils to groundwater is low. Parasites are larger and heavier than bacteria and viruses and do not readily move through soils.

Most of the early studies concerned pathogens in sewage sludge, biosolids, and wastewater. Since most biosolids applications are immediately incorporated into the soil, the potential for contamination of surface runoff is minimal. The low survivability of pathogens in soils also reduces the potential for surface water contamination.

Today, most biosolids are incorporated into the soil before a crop is planted. An exception is pastures, where liquid biosolids may be applied to an existing crop. Pathogens survive on plants for short periods of time as they are exposed to sunlight and ultraviolet light. Desiccation is also a major factor that reduces their survival.

The major danger to humans from soils in the food chain occurs primarily in developing countries where sanitation is poor. Human excrement is deposited in the field, often above water resources such as wells or streams.

REFERENCES

Alexander M, Wagenet RJ, Baveye RC, Gannon JT, Mingelgrin U, and Tan Y. 1991. *Movement of Bacteria through Soil and Aquifer Sand*. EPA/600/S2–91/010. Ada, OK: US Environmental Protection Agency, Robert S. Kerr Environmental Laboratory.

Andrews DA, Mawer SL, and Matthews PJ. 1983. Survival of salmonellae in sewage sludge injected into soil. *J Effluent Water Treat* (GB) 23: 72–74.

Babaeva RI.1966. Survival of beef tapeworm oncospheres on the surface of the soil in Samarkand. *Med Parazitiol Parazit Bolezen* 35: 557–560.

Bagdasaryan GA. 1964. Survival of viruses of the enterovirus group (poliomyelitis, ECHO, and coxsackie) in soil and on vegetables. *J Hyg Epidemiol Microbiol Immunol* 7: 497–505.

Beard PJ. 1938. The survival of typhoid in nature. *J Am Water Works Assoc* 30: 124.

Beard PJ. 1940. Longevity of *Eberthella typhosa* in various soils. *Am J Public Health* 30: 1077–1082.

Bicknell SR. 1972. *Salmonella aberdeen* in cattle associated with human sewage. *J Hyg* 70: 121–126.

Bitton G. 1975. Adsorption of viruses onto surfaces in soil and water. *Water Res* 9: 473–484.

Bitton G, Pancorbo OC, and Farrah SR. 1984. Virus transport and survival after land application of sewage sludge. *J Appl Environ Microbiol* 47(5): 905–909.

Brandl MT. 2006. Fitness of human enteric pathogens on plants and implications for food safety. *Annu Rev Phytopathol* 44: 367–392.

Bryanskaya AM. 1966. Antagonistic effect of Actinomyces on pathogenic bacteria in soil. *Hyg Sanit* 31: 123–125.

Burge WD and Enkiri NK. 1978. Virus adsorption by five soils. *J Environ Qual* 7: 73–76.

Butler RG, Orlob GT, and McGauhey PH. 1954. Underground movement of bacterial and chemical pollutants. *J Am Water Works Assoc* 46: 97–111.

Carlson JFJ, Woodward FE, Wentworth DF, and Sprou OJ. 1968. Virus inactivation on clay particles in natural waters. *Am J Public Health* 32: 1256–1262.

Casson LW. 1996. Fate and densities of pathogens in biosolids—what are our concerns? In *Water Environment Federation, 69th Annual Conference and Exposition*, Dallas TX, October 1996. Alexandria, VA: Water Environment Federation, 35–36.

Damgaard-Larsen S, Jensen KO, Lund E, and Nissen B. 1977. Survival and movement of enterovirus in connection with land disposal of sludges. *Water Res* 11: 503–508.

Drewry WA and Eliassen R. 1968. Virus movement in groundwater. *J Water Pollut Control Fed* 40: R257–R271.

Dunlop SG. 1968. Survival of pathogens and related disease hazards. In *Municipal Sewage Effluent for Irrigation*, ed. Wilson CW, and Beckett FE. Ruston, LA: Louisiana Tech Alumni Foundation, Tech. Station, 107–122.

Dunlop SG, Twedt RM, and Wang WL. 1951. *Salmonella* in irrigation water. *Sewage Ind Waste* 23: 118–122.

Dunlop SG and Wang WL. 1961. Studies on the use of sewage effluent for irrigation of truck crops. *J Milk Food Technol* 24: 44–47.

Epstein E. 1997. *The Science of Composting*. Boca Raton, FL: CRC Press.

Evans DJ, Casson LW, Sorber CA, Cockley KC, Keleti G, and Sagil BP. 1995. The transport and survivability of selected microorganisms in sludge-amended soils via rainfall-runoff studies. In *Residuals and Biosolids Management, Proceedings WEF 68th Annual Conference and Exposition*. Miami Beach, FL: Water Environment Federation, 2: 661–669.

Farrah SR, Scheuerman PR, and Bitton B. 1981. Urea-lysine method for recovery of enteroviruses from sludge. *Appl Environ Microbiol* 41: 459–465.

Feachem RG, Bradley DJ, Carelick H, and Mara DD. 1950. *Appropriate Technology for Water Supply and Sanitation: Health Aspects of Excreta and Silage Management—A State of the Art Review*. Washington, DC: World Bank.

Feachem JB, Bradley DJ, Garelick H, and Mara DD. 1980. *Appropriate Technology for Water Supply and Sanitation*. Washington, DC: World Bank.

Gerba CP. 1983. Pathogens. In *Utilization of Municipal Wastewater and Sludge on Land*, ed. Page AL, Gleason TLI, Smith JEJ, Iskandar K, and Sommers LE. Riverside: University of California, 147–185.

Gerba CP, Wallis C, and Melnick JL. 1975. Fate of wastewater bacteria and viruses in soil. *ASCE J Irrigation Drainage Div* 10: 157–174.

Gibbs RA, Hu CJ, Ho GE, and Unkovich I. 1997. Regrowth of faecal coliforms and salmonellae in stored biosolids and soil amended with biosolids. *Water Sci Technol* 35(11): 269.

Gilbert RG, Gerba CP, Rice RC, Bouwer H, Wallis C, and Melnick JL. 1976. Virus and bacteria removal from wastewater by land treatment. *Appl Environ Microbiol* 32: 333–338.

Gilbert RG, Rice RC, Bower H, Gerba CP, Wallis C, and Melnick JL. 1976. Wastewater renovation and reuse. Virus removal by soil filtration. *Science* 192: 1004–1005.

Jones F, Godfree AF, Rhodes P, and Watson DC. 1983. Salmonellae and sewage sludge—microbiological monitoring, standards and control in disposing sludge to agricultural lands. In *Biological Health Risks of Sludge Disposal to Land in Cold Regions*, ed. Wallis PM, and Lehmann DL. Calgary, Canada: University of Calgary Press, 95–114.

Kenner BA, Dotson GK, and Smith JE. 1971. *Simultaneous Quantitation of Salmonella Species and Pseudomonas aeruginosa II. Persistence of Pathogens in Sludge Treated Soils*. PB-213 706, National Tech. Info. Series. Washington, DC: US Environmental Protection Agency.

Kligler IJ. 1921. *Investigations of Soil Pollution and the Relation of the Various Privies to the Spread of Intestinal Infections*. Monograph 15. New York: International Health Board, Rockefeller Institute of Medical Research.

Kowal NE. 1982. *Health Effects of Land Treatment: Microbiological*. Report No. EPA/600/1-82-007. Cincinnati, OH: USEPA Health Effect Research Laboratory.

Lance JC. 1977. Fate of pathogens in saturated and unsaturated soils. In National Conference on Composting of Municipal Residues and Sludges. Silver Spring, MD: Information Transfer and Hazardous Materials Control Research Institute.

Larkin EP, Tierney JT, Lovett J, Van Dorsal D, and Francis DW. 1978. Land application of sewage wastes: potential for contamination of foodstuffs and agricultural soil by viruses. In *Risk Assessment and Health Effects of Land Application of Municipal Wastewater and Sludges*, ed. Sagic BP, and Sorber CA. San Antonio: University of Texas at San Antonio, 102–115.

Little MD. 1980. Agents of health significance: parasites. In *Sludge—Health Risks of Land Application*, ed. Bitton G, Damron BL, Edds GT, and Davidson JM. Ann Arbor, MI: Ann Arbor Science, 47–58.

Lue-Hing C, Sedita SJ, and Rao KC. 1979. Viral and bacterial levels resulting from land application of digested sludge. In *Utilization of Municipal Sewage Effluent and Sludge on Forest and Disturbed Land*, ed. Sopper WE, and Kerr SN. College Park, PA: University Press, 445–462

McGauhey PH and Krone RB. 1967. *Soil Mantle as a Wastewater Treatment System*, Report No. 67-11. Berkeley: University of California, Sanitary Engineering Research Laboratory.

Melick CO. 1917. The possibility of typhoid infection through vegetables. *J Infect Dis* 21: 28.

Moore BE, Sagic BP, and Sorber CA. 1978. Land application of sludges. Minimizing the impact of viruses on water resources. Proceedings of the Conference on Risk Assessment and Health Effects of Land Application of Municipal Wastewater and Sludges, San Antonio, TX.

Palfi A. 1972. Survival of enteroviruses during anaerobic digestion. In *Advances in Water Pollution Research. Proceedings of the Sixth International Conference, Jerusalem, Israel*, ed. Jenkins SH. New York: Pergamon Press.

Pepper IL, Josephson KL, Bailey RL, Burr MD, and Gerba CP. 1993. Survival of indicator organisms in Sonoran Desert soil amended with sewage sludge. *J Environ Sci Health* A28: 1287–1302.

Reddy KR, Khaleel R, and Overcash MR. 1981. Behavior and transport of microbial pathogens and indicator organisms in soils treated with organic wastes. *J Environ Qual* 10(3): 255–266.

Romero JC. 1970. The movement of bacteria and viruses through porous media. *Groundwater* 8: 37–48.

Rudolfs W, Falk LL, and Ragotzkie RA. 1950. Literature review of the occurrence and survival of enteric pathogenic, and relative organisms in soil, water, sewage, and sludge and on vegetation. *Sewage Indust Wastes* 22: 1261–1281.

Sepp E. 1971. *The Use of Sewage for Irrigation—A Literature Review*. Sacramento: State of California Department of Public Health, Bureau of Sanitary Engineering.

Shaub SA, Merer EP, Kolmer JR, and Sorber CA. 1975. *Land Application of Wastewater: The Fate of Viruses, Bacteria, and Heavy Metals at a Rapid Infiltration Site*. Report No. AD-A011263. Washington, DC: NTIS.

Sorber CA and Moore BE. 1986. Survival and transport of pathogens in sludge-amended soil. In *Proceedings of the National Conference on Municipal Treatment Plant Sludge Management*. Orlando, FL: Information Transfer, 25–32.

Stedman's. 1977. *Stedman's Medical Dictionary*, 23rd ed. Baltimore: Williams and Wilkins.

Straub TM, Pepper IL, and Gerba CP. 1992. Persistence of viruses in desert soils amended with anaerobically digested sewage sludge. *Appl Environ Microbiol* 58: 636–641.

Strauch D, Konig W, and Evers FH. 1981. Survival of salmonellae and *Ascaris* eggs during sludge utilization in forestry. In *Characterization Treatment and Use of Sewage Sludge*, ed. L'Hermite P, and Ott H. London Reidel, 408–416.

Theis JH, Bolton V, and Storm DR. 1978. Helminth ova in soil and sludge from twelve US urban areas. *J Water Pollut Control Fed* 50: 2485–2493.

Tierney JT, Sullivan R, and Larkin E. 1977. Persistence of poliovirus 1 in soil and on vegetables grown in soils previously flooded with inoculated sewage sludge or effluent. *Appl Environ Microbiol* 33: 109–113.

Van Dorsal DJ, Geldreich EE, and Clarke NA. 1967. Seasonal variations in survival of indicator bacteria in soil and their contribution to storm-water pollution. *Appl Microbiol* 15: 1362–1370.

Watson DC. 1980. The survival of salmonellae in sewage sludge applied to arable land. *J Water Pollution Control* 79: 11–18.

7 Geophagy and Human Pathogens in Plants

INTRODUCTION

In Chapter 6, I discussed the pathogens in soil and their impact on human health. This chapter presents another aspect of human health related to soils: geophagy. Geophagy is the deliberate ingestion of soil; the practice of consuming earthy or soil-like substances. Another term often used is pica (Izugbara 2003). Pica differs from geophagy in that it is less specific because it connotes the ingestion and appetite for largely nonnutritive substances. The Centers for Disease Control and Prevention (CDC), a unit of the National Institutes of Health, defines soil pica as eating 500 mg to more than 50 g of soil per day.

Soil and plants as related to pathogens and human health are interrelated. There are different variations of pica. Pica children in poor neighborhoods may ingest a soil near their residence. This soil, for example, could be contaminated by lead as a result of earlier lead-based paints or from tetraethyl lead-enhanced gasoline (Sing and Sing 2010). As an example, a young mother in Massachusetts received an old home when her grandmother died. The mother scraped the old paint from the home and refinished cabinets in the kitchen. Both her child and her dog came down with lead poisoning. Both had to undergo chelation, an unpleasant procedure. Removing the soil surrounding the house was also extremely expensive as Massachusetts required that the contaminated soil removed be sent to a hazardous waste site. The alternative was to bind the lead with organic matter and phosphorus and to find a place on the home grounds that was free of lead, which then became the child's playground.

GEOPHAGY

Geophagy has been recorded in every region of the world and is reported as a pathological condition by those in the medical profession (Reid 1992). It is most prevalent in the world's poorest regions (Abrahams 1997; Abrahams and Parsons 1997). Although there are some nutritional aspects to soil ingestion or geophagy, the purpose of this section is to point out the potential health hazards, excluding nonpathogens.

Generally, pregnant women and children are the greatest consumers (Callahan 2003). Childhood geophagy routinely involves topsoil (upper 15 cm). Geophagy has both positive and negative aspects. The positive aspects most reported in the literature refer to nutritional benefits. The practice has been observed in many cultures, primarily in pregnant women (Njiru, Elchalal, and Paltiel 2011; Sing and Sing 2010). Sing and Sing indicate that colloidal clays are consumed for antidiarrheal properties, and that geophagy is practiced for nutritive substances where normal diets do not have the full benefits of vegetative nutrients. Another aspect is the provision of calcium, which is also associated with low supply in certain regions (Sing and Sing 2010). In Bangladesh, where there is a high concentration of arsenic in drinking water, an additional source of arsenic could be soil ingestion. Further, baked clay, which is ingested, may also contain cadmium and lead as well as arsenic (Al-Rmalli et al. 2010).

One aspect that has been pointed out is cultural (Izugbara 2003).

The practice is primarily among children of both sexes, school-aged children, teenage mothers, and women. There are few studies in this regard. A summary of a study involving pregnant women in southeastern Nigeria is shown in Table 7.1. The study of geophagy was conducted by the University of Uyo and Abia State University in Nigeria. Table 7.2 shows the frequency of geophagy and types of soil consumed.

TABLE 7.1
Sociodemographic Characteristics of Southeastern Nigeria Pregnant Persons Associated with Geophagy

Pregnant	Percentage
Age (years)	
Less than 30	31.3
31–40	42.1
41 and older	27
Highest level of education	
No formal education	33.3
Primary (1–6 years)	49.6
Secondary	14.6
Beyond 14 years of age	2.5
Marital status	
Single	2.5
Married	90
Religion	
Catholic	39.6
Protestant	58.3
Other	2.1
Occupation	
Housekeeper	89.6
Farmer	97.5
Petty trader (businesswomen)	82.5

Source: Modified from Izugbara CO. 2003. The cultural context of geophagy among pregnant and lactating Ngwa women of southeastern Nigeria. *Afr Anthropol* 10: 180–199.

TABLE 7.2
Frequency of Geophagy and Types of Soil Consumed

Frequency of Soil Eaten	Percentage
Eats soil at least once per day	78.3
Eats soil less than seven times per week	15
Does not eat soil at all	67
Types of soil eaten	
Clay	93
Soil from gullies and paths	27
Walls of huts	59.6
Soil from cooking mounds	65.4
Termite mounds	64.2

Source: Modified from Izugbara CO. 2003. The cultural context of geophagy among pregnant and lactating Ngwa women of southeastern Nigeria. *Afr Anthropol* 10: 180–199.

Although this was a sociodemographic study, two aspects would have been interesting: the chemical content of the clay/soil and any frequency of infection from parasites and other bacteria or viruses.

Another aspect as a result of ingesting soil, especially in developing countries, is the risk from soil contaminated by human and animal feces containing parasites. Two children at two different locations were infected with raccoon roundworm (*Baylisascaris procyonis*) as a result of ingesting soil (Callahan 2003; CDC 2002).

Although most of the recent literature on geophagy is from incidences in developing countries, Flynt's *Dixie's Forgotten People: The South's Poor Whites* indicates that geophagy was common among poor whites (Flynt 2004). The CDC indicates that the most common infection in the United States is associated with parasitic infection caused by *Toxocara canis*, and the most common route of infection is contact with soil contaminated with cat and dog feces. The parasite can remain in the soil for years (Callahan 2003).

Vermeer and Frate (1979) reported on an investigation of the black population in rural Holmes County in Mississippi. They found that geophagy occurred among 57% of women and 16% of children of both sexes but was not found among male adults or adolescents. The data did not correlate with hunger, anemia, or parasitic problems (Vermeer and Frate 1979). Geophagy has also been associated with mineral deficiency. In a study in Turkey, iron and zinc deficiencies were found among Turkish children and women in villages (Cavdar et al. 1983). Several symptoms were observed, including anemia, growth retardation, and hypogonadism. Hooda et al. (2002) evaluated five geophagic materials from Uganda, Tanzania, Turkey, and India with mineral nutrient concentrations and conditions similar to the gastrointestinal tract. They reported that while calcareous geophagic materials may supplement calcium, geophagy could result in iron and zinc deficiency. Dreyer, Chaushev, and Gledhill (2004) indicate that the local African population believed that ingestion of soil benefits pregnant women suffering from iron deficiency.

The predominant literature regarding geophagy relates to human infections from organisms. As would be expected, infections from geophagy related to diseases are more common in developing countries. In 1998, Geissler et al., in a study of 204 children aged 10–18 in western Kenya, found infections by the helminth *Ascaris lumbricoides, Trichuris trichiura,* and *Schistosoma mansoni.* There were significant associations between geophagy and infection intensity with *Ascaris lumbricoides* and *Trichuris trichiura* (Geissler et al. 1998*).*

There were also significant associations between geophagy and *Ascaris lumbricoides* and not hookworm, *Trichuris trichiura,* and *Strongyloides stercoralis* and high microbial viable counts (maximum 120,000 CFU/g) (Kutalek et al. 2010). Similar results were reported by Kawai et al. (2009).

One aspect rarely reported is the relation of airborne dust and geophagy to human health. Airborne dust can carry not only spores but also fungi, endotoxins, bacteria, and a large variety of microorganisms. Sing and Sing (2010) point out the difficulty in studying dust components. Data from several US areas as well as Mali, Israel, Virgin Islands, and Korea are shown in Table 7.3. Although not enumerated, amounts of bacteria in dusts were high. Undoubtedly, several of the bacteria could have been pathogenic. Dust containing fungal spores could increase asthmatic symptoms and can affect respiratory conditions.

It is evident that supplemental multivitamins given to residents in villages of developing countries can overcome some of the health aspects resulting from infections due to geophagy that are much more costly and difficult to eradicate. This action must be coupled with education and appropriate sanitary conditions.

TABLE 7.3
Bacterial and Fungal Concentrations in Dust Storms

Sample Location	Dust Source	Bacterial Background (CFU/m³)	Bacteria in Dust (CFU/m³)	Fungal Concentrations in Background (CFU/m³)	Fungal Concentrations in Dust (CFU/m³)
Kansas	Kansas	<10	2,880–42,735	Not detected	Not detected
Junction	Texas	<450	>1,544	Not applicable	Not applicable
Mali	Sahara/Sahil	200–1,100	720–15,700	0–130	80–370
Israel	Sahara	79–108	694–995	31–115	205–226
US Virgin Islands	Sahara/Sahil	0–100	90–350	0–60	30–60
Korea	Gobi/Taklamakanm	105–1,930	225–8,212	100–8,510	336–6,992

Source: Modified from Sing D, and Sing CF. 2010. Impact of direct soil exposure from airborne dust and geophagy on human health. *Int J Environ Res Public Health* 7: 1205–1223.

Note: CFU/m³, colony-forming units per cubic meter.

PATHOGENS ON PLANTS

Soil, plants, and human pathogens are interrelated. Normally, we associate human diseases from pathogens and plants as the result of contamination. Contamination can be caused by improper worker sanitation as well as contamination of produce from wild or domestic animals or contaminated wastewater. Workers with proper education and incentives can be taught to identify potential sources of contamination. Wild and domestic animals entering a produce field can contaminate the produce through defecation. The improper use of manures that have not been sterilized (e.g., compost that has not reached 55°C) can also be a source of contamination. In many cases, the soil becomes contaminated, resulting in contamination of the food produced.

However, earlier data (Table 7.4) show that once human pathogens infect plants, they can survive for long periods of time. More recently, it has been shown that there is a relationship between human pathogens and the rhizosphere. Organisms associated with plants can be fungi, bacteria, and viruses.

In recent years, there have been numerous incidents of foodborne diseases as a result of contamination of food crops. This could be the result of pathogen survival in soils and contaminating crops. Manure, which often is untreated for destruction of pathogens, could result in contamination as could irrigation waters. There was also an indication that wild animals entering food-growing fields could result in contamination of crops. A schematic in Figure 7.1 shows the factors that can contribute to the contamination of plants (fruits and vegetables) with human enteric pathogens in the field (Brandl 2006). Brandl indicated that the first step in the contamination of the aerial portion of the plant is attachment. The author provided an excellent review. There are several potential sources of contamination of field-grown fruits

TABLE 7.4
Survival of Pathogens or Indicator Organisms on Plants

Organism	Plant	Survival Time	References
Poliovirus 1	Lettuce Radish	23 days	Tierney, Sullivan, and Larkin 1977
Coliforms	Tomatoes	35 days	Englebrecht 1978
Salmonella typhi	Vegetables	7–53 days	Englebrecht 1978
Salmonella typhi	Vegetables and fruit	<1–68 days	Parsons et al. 1975
Shigella spp.	Vegetables	2–10 days	Parsons et al. 1975
Shigella spp.	Tomatoes	2–7 days	Englebrecht 1978
Vibrio cholerae	Vegetables	2 days	Englebrecht 1978
Tubercle bacilli	Radish	90 days	Englebrecht 1978
Tubercle bacilli	Grass	10–49 days	Parsons et al. 1975
Entamoeba histolytica	Vegetables	3 days	Englebrecht 1978
Entamoeba histolytica	Vegetables	<1–3 days	Parsons et al. 1975
Taenia saginata eggs	Pasture	90–365 days	Englebrecht 1978
Enteroviruses	Vegetables and fruits	4–6 days	Parsons et al. 1975

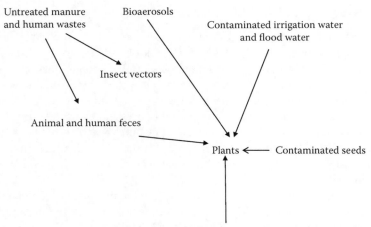

FIGURE 7.1 Pathway of plant contamination by human pathogens.

and vegetables. The application of untreated (raw) manure to fields can result in enteric pathogens in the fields (Solomon, Brandl, and Mandrell 2006). Thus, the use of untreated or unstabilized manure can result in the contamination of food crops. Furthermore, organisms from a manure pile or field where manure has been applied can contaminate irrigated water or bodies of water used for swimming or drinking. This also applies to free-roaming domestic or wild animals, whose fecal matter can contaminate food or water resources. Many immigrant workers do not have proper sanitation conditions or knowledge regarding the importance of sanitation and can transmit disease to other workers or members of the family as well as the crops harvested. *Escherichia coli* O157:H7 and other enteric bacteria can contaminate the surface of edible plants both pre- and postharvest (Aruscavage, Miller, and LeJeune 2006). There are other reported incidents of pathogenic bacteria and viruses that can attach to fruits and vegetables. These soil organisms have been *Listeria monocytogenes* and *Clostridium botulinum*.

According to earlier reports, Rudolfs, Falk, and Ragotzkie (1950) and Bryan (1977) indicated that pathogens do not penetrate into fruits or vegetables unless their skin is broken. Bryan reviewed the early literature on the survival of pathogens on crops. Pathogen survival on fruits and vegetables exposed to sunlight and drying would be short. The survival on subsurface crops such as potatoes and beets would be similar to those in soil (Gerba 1983).

Crops whose edible portion does not come in contact with the soil or biosolids are less apt to be contaminated. This is the basis for the Part 503 regulations concerning class B biosolids used for land application. Table 6.6 provides information on the survival of indicator organisms and pathogens on plants.

Kowal (1982) indicated that survival times of several bacterial pathogens ranged from less than 1 day to 6 weeks. Virus survival on the surface of aerial crops would be expected to be shorter than in soil because of exposure to deleterious environmental effects, especially sunlight, high temperatures, drying, and washing off by rainfall (Kowal 1982). Gerba (1983) indicated that the intact surfaces of vegetables are probably impenetrable for viruses. Parasites have been reported to survive on plant surfaces for months. Sepp (1971) reported that *Ascaris* ova were destroyed in 27–35 days on vegetable surfaces during dry summer weather by desiccation. In a study by Ohio University and the Ohio Farm Bureau federation in Ohio (US Environmental Protection Agency [USEPA] 1985), soil and forage samples were collected for parasitic ova and larvae on three farms. Samples were taken from pasture lands treated with biosolids and untreated pasture lands. The analysis was done before biosolid application and 7, 14, and 28 days following application. The study concluded that the risk of parasite transmission to animals from land with applied biosolids was indistinguishable from farms without such application.

Gagliardi and Karns (2002) studied the persistence of *E. coli* O157:H7 in fallow and manured soil and plants. They reported that the bacteria persisted for 25–41 days on rye roots and for 47–96 days on alfalfa roots. Islam et al. (2005) found that they survived for 154–196 days on carrots and onions. Walker et al. (2004) indicated that *Pseudomonas aeruginosa*, an opportunistic human pathogen, is capable of infecting sweet basil and *Arabidopsis*.

The antimicrobial activity of some medicinal plant extracts have been shown to inhibit antibacterial activity against *Bacillus subtilis, E. coli, Pseudomonas fluorescens, Staphylococcus aureus, Xanthomonas axonopodis* pv. *malvacearum*, and *Aspergillus flavus* (Mahesh and Satish 2008). As plants are continuously in contact with microorganisms (i.e., bacteria, fungi, and viruses), they can accumulate antimicrobial secondary metabolites (Gonzalez-Lamothe et al. 2009).

It is evident from the literature that the survival rate of pathogens on plants is short. Desiccation, temperature, and ultraviolet light are the most important factors in destroying pathogens on plants. Thus, the risk for humans consuming foods grown where biosolids have been applied to the land is even lower since most of the biosolids are incorporated into the soil and do not come in contact with edible food crops. The risk to humans from pathogens in biosolids that are applied to nonedible crops, forestry, and fruit trees is essentially nil.

Aruscavage, Miller, and LeJeune (2006) state that pathogens on edible plants present a significant potential source of human illness. Even though some pathogens do not survive on plant surfaces or are removed by washing, a significant portion of enteric pathogens can remain in association with fresh produce (Aruscavage, Miller, and LeJeune 2006). As these authors report, contrary to conventional wisdom, *Escherichia coli* and *Salmonella* survive outside the animal host. Often, irrigation water was insufficient to remove microorganisms (Brandl 2006). In 2006, increased foodborne disease outbreaks in the United States were due primarily to *E. coli* O157:H7, *Listeria monocytogenes, Salmonella* serotype Enteritidis, and *Clostridium botulinum* (CDC 2009). It is not known how the opportunistic fungal pathogen *Cryptococcus* can complete its sexual cycle in association with plants (Xue et al. 2007).

Tyler and Triplett (2008) showed in laboratory studies that enteric human pathogens such as *Salmonella* and *E. coli* O157:H7 can enter plants. Plant defenses can regulate the extent of entry.

The removal of enteric pathogens during washing and sanitizing may not be effective if the organism is localized on subsurface sites on leafy green plant tissue (Hirneisen, Sharma, and Kniel 2012). Akhtyamova (2013a, 2013b) also indicated that members of the family Enterobacteriaceae, including pathogenic *Salmonella* and *Shigella* strains, *Vibrio cholerae*, and *Pseudomonas aeruginosa* were found in plants or inside plants.

The survival time of several pathogenic or indicator organisms on plants are shown in Table 7.4. Most did not survive for long periods. However, tubercle bacilli did survive for 90 days.

It is evident that the contamination and survival of human pathogenic organisms on plants can result in human diseases. This can be minimized by good hygienic practices and the decontamination of fruits and vegetables by washing.

SUMMARY

This chapter covered two topics related to Chapter 6, Pathogens in Soil. In this chapter, one of the less-known areas was geophagy—the deliberate ingestion of soil. There are two aspects to this condition. Ingestion of soil can be beneficial through the addition of minerals such as calcium, iron, and zinc and deleterious by ingestion of human pathogens.

Earlier literature indicates that geophagy occurred in poor and less-developed countries as a nutritive measure that provided iron, calcium, and zinc to pregnant mothers and children. What was ignored was the harboring of human pathogens in the soil as a result of human defecation. In many villages in developing countries, children and adults defecate outside the home and do not properly discharge their feces. Subsequently, children ingesting contaminated soil will ingest pathogens, resulting in illnesses such as diarrhea.

This chapter also discussed pathogens on plants, which is further elaborated in the chapter on foodborne pathogens and diseases. Pathogens on plants generally do not survive for long periods of time because of exposure to ultraviolet light, temperature, and desiccation. However, where food crops are produced, the short harvest period and further contamination of other plants can result in human health issues.

REFERENCES

Abrahams PW. 1997. Geophagy (soil consumption) and iron supplementation in Uganda. *Trop Med Int Health* 2: 617–623.

Abrahams PW, and Parsons JA. 1997. Geophagy in tropics: an appraisal of three geophagical materials. *Environ Geochem Health* 19: 0.

Akhtyamova N. 2013a. Human pathogens on or within the plant and useful endophytes. *Cell Dev Biol* 2: e110.

Akhtyamova N. 2013b. Human pathogens—the plant and useful endophytes. *Med Microbiol Diagn* 2: e121.

Al-Rmalli SW, Jenkins RO, Watts MJ, and Haris PI. 2010. Risk of human exposure to arsenic and other toxic elements from geophagy: trace element analysis of baked clay using inductively coupled plasma mass spectrometry. *Environ Health* 9: 79.

Aruscavage KL, Miller S, and LeJeune TJ. 2006. Interactions affecting the proliferation and control of human pathogens on edible plants. *J Food Sci* 71: R88–R99.

Brandl MT. 2006. Fitness of human enteric pathogens on plants and implications for food safety. *Annu Rev Phytopathol* 44: 367–392.

Bryan FL. 1977. Disease transmitted by foods contaminated by wastewater. *J Food Prot* 40: 45–52.

Callahan GN. 2003. Eating dirt. *Emerg Infect Dis* 9: 1016–1021.

Cavdar AO, Arcasoy A, Cin S, Babacan E, and Gozdasoglu S. 1983. Geophagia in Turkey: iron and zinc deficiency, iron and zinc absorption and studies and response to treatment with zinc in geophagia cases. *Prog Clin Biol Res* 129: 71–97.

Centers for Disease Control and Prevention (CDC). 2002. Raccoon roundworm encephalitis—Chicago, Illinois and Los Angeles, California. *MMWR Morb Mortal Wkly Rep* 50(51): 1153–1155.

Centers for Disease Control and Prevention (CDC). 2009. Surveillance for foodborne disease outbreaks—United States, 2006. *MMWR Morb Mortal Wkly Rep* 58(22): 609–615.

Dreyer MJ, Chaushev PG, and Gledhill RF. 2004. Biochemical investigations in geophagia. *J R Soc Med* 97: 48.

Englebrecht RS. 1978. Microbial Hazards Associated with the Land Application of Waste Water and Sludge. Ernest Balsom Lecture, University of London.

Flynt W. 2004. *Dixie's Forgotten People: The South's Poor Whites.* Bloomington: Indiana University Press.

Gagliardi JV, and Karns JS. 2002. Persistence of *Escherichia coli* O157:H7 in soil and on plant roots. *Environ Microbiol* 4: 89–96.

Geissler PD, Mwaniki D, Throng F, and Friis H. 1998. Geophagy as a risk factor for geohelminths infections: a longitudinal study of Kenya primary schoolchildren. *Trans R Soc Trop Med Hyg* 92: 7–11.

Gerba CP. 1983. Pathogens. In *Utilization of Municipal Wastewater and Sludge on Land*, ed. Page AL, Gleason TLI, Smith JEJ, Iskandar IK, Sommers LE. Riverside: University of California, pages 147–185.

Gonzalez-Lamothe R, Mitchell G, Gattuso M, Diarra MS, Malouin F, and Bouarab K. 2009. Plant antimicrobial agents and their effects on plant and human pathogens. *Int J Mol Sci* 10: 3400–3419.

Hirneisen KA, Sharma M, and Kniel KE. 2012. Human enteric pathogen internalization by root uptake into food crops. *Foodborne Pathog Dis* 9: 369–405.

Hooda PS, Henry CJK, Seyoum TA, Armstrong LDM, and Fowler MB. 2002. The potential impact of geophagia on the bioavailability of iron, zinc and calcium in human nutrition. *Environ Geochem Health* 24: 305–319.

Islam M, Doyle M, Phatak SC, Millner P, and Jiamg X. 2005. Survival of *Escherichia coli* O157:H7 in soil, and on carrots and onions grown in fields treated with contaminated manure composts or irrigation water. *Food Microbiol* 22: 63–70.

Izugbara CO. 2003. The cultural context of geophagy among pregnant and lactating Ngwa women of southeastern Nigeria. *Afr Anthropol* 10: 180–199.

Kawai K, Saathoff E, Antelman G, Msamanga G, and Fawzi WW. 2009. Geophagy (soil-eating) in relation to anemia and helminth infection among HIV-infected pregnant women in Tanzania. *Am J Trop Med Hyg* 8: 36–43.

Kowal NE. 1982. *Health Effects of Land Treatment: Microbiological.* Report no. EPA/600/1-82-007.Cincinnati, OH: USEPA Health Effect Research Lab.

Kutalek R, Wewalka G, Gundacker C, Auer H, Wilson J, Haluza D, Huhulescu S, Hiller S, Sager M, and Prinz A. 2010. Geophagy and potential health implications: geohelminths, microbes and heavy metals. *Trans R Soc Trop Med Hyg* 104: 787–795.

Mahesh B and Satish S. 2008. Antimicrobial activity of some important medicinal plant against plant and human pathogens. *World J Agric Sci* 4: 839–843.

Njiru H, Elchalal U, and Paltiel O. 2011. Geophagia during pregnancy in Africa: a literature review. *Obstet Gynecol Surv* 66: 452–459.

Parsons HR, Brownlee C, Wetter D, Maurer A, Haughton E, Kordner L, Slezak M. 1975. *Health Aspects of Sewage Effluent Irrigation*. Victoria, BC: Pollution Control Branch, British Colombia Water Resources Service.

Reid RM. 1992. Cultural and medical perspectives on geophagia. *Med Anthropol* 13: 337–351.

Rudolfs W, Falk LL, Ragotzkie RA. 1950. Literature review of the occurrence and survival of enteric pathogenic, and relative organisms in soil, water, sewage, and sludge and on vegetation. *Sewage Indust Wastes* 22: 1261–1281.

Sepp E. 1971. *The Use of Sewage for Irrigation—A Literature Review*. Sacramento: State of California Department of Public Health, Bureau of Sanitary Engineering.

Sing D and Sing CF. 2010. Impact of direct soil exposure from airborne dust and geophagy on human health. *Int J Environ Res Public Health* 7: 1205–1223.

Solomon EB, Brandl MT, and Mandrell RE. 2006. Biology of food-borne pathogens on produce. In *Microbiology of Fresh Produce*, ed. Matthews KR. Washington, DC: ASM Press, 267–391.

Tierney JT, Sullivan R, Larkin E. 1977. Persistence of poliovirus 1 in soil and on vegetables grown in soils previously flooded with inoculated sewage sludge or effluent. *Appl Environ Microbiol* 33: 109–113.

Tyler HL and Triplett EW. 2008. Plants as a habitat for beneficial and/or human pathogenic bacteria. *Annu Rev Phytopathol* 46: 53–73.

US Environmental Protection Agency (USEPA). 1985. *Demonstration of Acceptable Systems for Land Disposal of Sewage Sludge*. Cincinnati, OH: Water Engineering Research, Office of Research and Development.

Vermeer DE and Frate DA. 1979. Geophobia in rural Mississippi: environmental and cultural contents and nutritional implications. *Am J Clin Nutr* 32: 2129–2135.

Walker TS, Bais HP, Deziel E, Schweizer HP, Rahme LC, Fall R, and Vivanco JM. 2004. *Pseudomonas aeruginosa*–plant root interactions. Pathogenicity, biofilm formation, and root exudation. *Plant Physiol* 134: 320–331.

Xue C, Tade Y, Dong X, and Heltman J. 2007. The human fungal pathogen *Cryptococcus* can complete its sexual cycle during a pathogenic association with plants. *Cell Host Microbe* 1: 263–273.

8 Bioaerosols

INTRODUCTION

Bioaerosols can contain living organisms, including bacteria, viruses, fungi, actinomycetes, arthropods, and protozoa. They can also contain microbial products such as endotoxin, microbial enzymes, β-1(3)-glucans, and mycotoxins (American Conference of Governmental Industrial Hygienists [ACGIH] 1999). Many of the bioaerosols are ubiquitous in both indoor and outdoor environments. The sources can be soil, plants, organic matter, animals, and humans.

Polymemkou (2012) indicates that the atmosphere has been one of the latest frontiers of biological exploration. Humans can obtain pathogens by inhalation of pathogenic airborne particles transmitted by other humans. Table 8.1 lists some of the diseases transmitted by humans and caused by inhalation. There are several other diseases that can result from inhalation of particles, as shown in Table 8.2.

The US Department of Defense has indicated that military personnel serving in camps located in southwestern Asia and Afghanistan have complained of exposure to dust in the air containing particles from fecal matter. These particles can result in eye and nasal irritation and other respiratory discomfort and health aspects. Furthermore, inadvertent contact with fecal matter and poor hygiene can result in severe infections.

Occupational exposure to bioaerosols occurs in numerous agricultural industries, as well as nonfarming practices. These can include the following:

- Activities related to animal manure
- Animal feeding activities
- Composting
- Pulp and paper industry
- Horticultural industry (e.g., greenhouses, landscaping, turf and sod production)
- Zoological activities
- Public works activities
- Veterinary activities

One aspect related to air pollution is asthma. Asthma is characterized by chronic inflammation of the airways. Breathing becomes difficult. An attack of asthma results in constricted and swollen airways filled with mucus. It is a complex condition. Numerous molecular immune system pathways are being explored. More details are

TABLE 8.1
Some Important Inhaled Diseases and Their Organisms

Disease	Pathogen
Chicken pox	*Varicella*
Flu	Influenza
Measles	Rubeola
German measles	Rubella
Mumps	Mumps
Smallpox	Variola
Whooping cough	*Bordetella pertussis*
Meningitis	*Neisseria*
Diphtheria	*Corynebacterium diphtheriae*
Pneumonia	*Mycoplasma pneumoniae*
Tuberculosis	*Mycobacterium tuberculosis*

Source: Modified from Deacon, date not available.

TABLE 8.2
Some Pathogens Not as Prevalent in the Air

Disease	Organism	Source
Psittacosis	*Chlamydia psittaci*	Bird fecal particles
Legionnaires' disease	*Legionella pneumophila*	Air conditioning systems
Acute allergic alveolitis	Fungal and actinomycetes spores	Decomposing organic matter
Aspergillosis	*Aspergillus fumigatus* and *Aspergillus niger*	Composting, decomposing organic matter
Histoplasmosis	*Histoplasma capsulatum*	Fungal spores from bird and bat fecal matter
Coccidioidomycosis	*Coccidioides immitis*	Spores in dust

Source: Modified from Deacon L, Parkhurst L, Liu J, Gewq GH, Hayes ET, Jackson S, Longhurst J, Longhurst P, Pollard S, and Tyrrel S. 2009. *Environ Health* 8: 51–59.

available in a recent article (Maxmen 2011). There are numerous conditions related to air quality resulting in asthma. These can be the following:

- Sulfur dioxide as a result of burning coal and crude oil.
- Ground-level ozone as a component of smog. This can trigger asthma. It is a major problem in cities with pollution from cars and trucks.
- Particulate matter, which is the result of numerous pollutants, such as dust, soot, fly ash, diesel exhaust particles, wood smoke, and aerosols.
- Nitrogen oxide, which is emitted from automobile tailpipes and power plants.

The World Health Organization (WHO) indicated in 2013 that some 234 million persons suffer from asthma. It is the most common chronic disease among children. Most asthma-related deaths occur in low- and lower-income countries. The strongest risk factors are inhaled substances and particles that may produce allergic reactions or irritation to the airways (WHO 2013). There are numerous factors that trigger asthma. These can include

- Dust
- Mites
- Pollen and molds
- Tobacco smoke
- Chemical irritants in the workplace
- Air pollution

Other factors can be psychological or related to air temperature and even certain medications. Although the role of human pathogens has not been cited as a direct source, bioaerosols can increase asthmatic conditions.

Bioaerosols are found both indoors and outdoors. Many occupations are in greater contact with bioaerosols; therefore, these workers are more vulnerable to the respiratory effects. The most noted such workers are

- Farmers involved with animal manure, animal feedstock, grain, hay, and silo material
- Composters and others involved with decaying organic matter and yard materials
- Workers dealing with municipal solid waste
- Workers involved with biosolids, especially in closed environments
- Workers in the pulp and paper industry, wood products, and lumber
- Employees in the horticultural industry, especially greenhouses, and in landscaping, turf, and sod production
- Workers in zoological gardens
- Public work employees in parks and recreational facilities
- Veterinarians

Sources of indoor bioaerosols have been identified as follows (Cox and Wathes 1995):

- Cooling towers
- Building exhausts
- Humidification systems
- Homes
- Hospitals
- "Sick buildings"
- Laboratories

TABLE 8.3
Fungal and Bacterial Concentrations in Several Industries

Industry	Fungi (CFU/m^2)	Bacteria (CFU/m^2)
Agricultural harvesting and storage	10^3–10^9	10^2–10^3
Animal facilities	10^2–10^8	10^3–10^5
Composting	10^2–10^7	10^3–10^7
Manufacturing technology	10^2–10^6	10^2–10^6
Sawmill	10^4–10^8	10–10^3
Wastewater treatment (activated sludge)	10–10^3	10^2–10^6

Source: Modified from Prasad M, van der Werf P, and Brickman A. 2004. Bioaerosols and composting—a literature evaluation. Composting Association of Ireland. http://www.cre.ie/docs/cre_bioaerosol_aug2004.pdf (accessed February 18, 2011).

Table 8.3 provides some data on concentration of fungi and bacteria in the air of several industries.

FUNGI AND PATHOGENS COMMONLY FOUND IN OUTDOOR AND INDOOR ENVIRONMENTS

The outdoor environment is the principal source of bioaerosols in the indoor environment. The main environmental areas are the natural environment; agricultural activities; wastewater treatment facilities; and solid waste activities, including handling, collection, and disposal.

Many of the indoor airborne fungi were identified by Levetin (n.d.). Some of these are listed next:

- *Acremonium*: Common species, found in soil, decaying vegetation and food sources. Several species have been associated with human diseases, such as meningitis, midline granuloma, and infections.
- *Alternaria*: Common outdoor fungus. Many times, it has been considered the second-most-common mold spore genus. Numerous infections have been reported. Lesions have been related to immunosuppressant conditions. The spores are highly allergenic.
- *Aspergillus* spp.: The environmental source is plant debris, decaying vegetation, organic matter, and soil. It is also found in household dust and construction material. There are over 185 species, and many cause opportunistic infections in humans. One species, *Aspergillus fumigatus*, has been considered a ubiquitous organism in the environment. There are many different species of *Aspergillus*, but the most common species are *Aspergillus fumigatus* and *Aspergillus flavus*. Other species are *Aspergillus terreus*, *Aspergillus nidulans*, and *Aspergillus niger*.

- *Basidiospores*: Spores produced by mushrooms, puffballs, shelf fungi, rusts, smuts, and many other fungi.
- *Chaetomium*: It is a genus of fungi. As a mold, it is found outdoors in the air and in soil and plant debris. It can produce an infection in humans.
- *Cladosporium*: It is a fungal genus found both indoors and outdoors. It is a weak plant pathogen. Indoors, it can be found in carpets and wallpaper. It is not known to be a serious human pathogen.
- *Curvularia*: It is an outdoor fungus or mold and a facultative pathogen of many plant species.
- *Drechslera*: An outdoor fungus whose species are often plant pathogens.
- *Epicoccum*: A saprophyte occurring both indoors and outdoors. It is essentially a plant pathogen.
- *Fusarium*: A common saprophyte in the soil and a plant pathogen. Some species produce mycotoxins in cereal crops that can affect human and animal health if they enter the food chain.
- *Myrothecium*: Soil fungus that can produce mycotoxins.
- *Paecilomyces*: It is a fungus occurring both indoors and outdoors and can form mycotoxins.
- *Penicillium*: It is a common fungal genus. It is found both indoors and outdoors. Some species can produce mycotoxins.

There are many other fungi that produce spores. Although they are in the outdoor air, they can also be found indoors in moist environments. Fungal molds are often found in water-damaged homes (Fabian et al. 2005).

Pathogenic fungi and their environmental sources are shown in Table 8.4. *Candida* spp., *Aspergillus* spp., and *Cryptococcus neoformans* are the most common pathogenic fungi. *Candida* spp. constitute the third to fourth most common cause of bloodstream infections occurring in hospitals or infirmaries (Walsh et al. 2004). Rhodes (2006) stated that *Aspergillus fumigatus* is the leading mold pathogen among immunocompromised patients, especially those who have had bone marrow and solid organ transplantation. *Cryptococcus neoformans* is the most common cause of fungal-related mortality in patients with human immunodeficiency virus (HIV) (Walsh et al. 2004). Walsh et al. indicate that *Fusarium* spp., *Scedosporium* spp., *Trichoderma* spp., Zygomycetes, *Penicillium marneffei*, and *Trichosporons* spp. are the most significant emerging fungi. Immunocompromised patients are the most vulnerable. One important issue is the presence of bioaerosols in hospital rooms (McGinnis et al. 2009).

Bioaerosol infections in hospitals have been reported (Nobel and Clayton 1963; Burge 2008; Mullins, Harvey, and Seaton 1976; Rodenhuis et al. 1984; Bodey and Vartibvarian 1989; Arnow et al. 1978; Summerbell, Krajden, and Kane 1989; Burton et al. 1972; Hopkins, Webber, and Rubin 1989). These and subsequent reports showed that nosocomial infections or secondary infections may be acquired from hospitals during treatment (Ekhaise, Ighosewe, and Ajakpori 2008). Ekhaise et al. reported on the concentration of bioaerosols in various hospital departments. The bacterial population ranged from 3.0 to 76 CFU/m^3. The highest levels were in the accident and emergency ward. The fungal population ranged from 6.0 to 44.7 CFU/m^3. The

TABLE 8.4
Some Pathogenic Fungi and Their Environmental Sources

Fungus	Disease	Environmental Source	Reference
Aspergillus spp.	Allergic bronchopulmonary aspergillosis; colonization; tissue invasion; toxicosis	Soil, plant debris, organic matter, compost air	Latgé 1999; Epstein 2011; Kwon-Chung and Bennett 1992
Candida spp. *Candida albicans*	Candidiasis; opportunistic mycosis; bloodstream infection	Gastrointestinal tract; leaves, flowers, soil, water	Dagnani, Solomkin, and Anaissie 2003; Epstein 2011; www.doctorfungus.org. 2006
Cryptococcus neoformans	Cryptococcosis	Soil contamination from avian excreta; pigeon droppings	Viviani, Tortorano, and Ajello 2003; Epstein 2011; www.doctorfungus.org. 2006
Fusarium spp. *Fusarium solani*	Plant pathogen; keratomycosis; mycetoma; onychomycosis; mycotoxicosis	Soil; decaying vegetation; air	Dagnani, Kiwan, and Anaissie 2003; Epstein 2011; www.doctorfungus.org. 2006
Penicillium spp. *Penicillium marneffei*	Penicilliosis; pulmonary infection	Soil; decaying vegetation; air	www.doctorfungus.org. 2006; Kwon-Chung and Bennett 1992; Epstein 2011
Scedosporium spp.	Sinusitis; brain abscess; meningitis; fungus ball; endocarditis	Soil; polluted water; compost	Epstein 2011; Kwon-Chung and Bennett 1992; Dagnani, Kiwan, and Anaissie 2003)
Trichoderma spp.	Peritonitis; infections in immunocompromised individuals	Soil; plant material; common house mold; decaying vegetation; wood	www.doctorfungus.org. 2006; Kwon-Chung and Bennett 1992; www.doctorfungus.org. 2006; Epstein 2011
Trichosporon	Trichosponosis; infections	Soil; water; birds; mammals; mouth; skin; nails; vegetables	Kwon-Chung and Bennett 1992; www.doctorfungus.org. 2006; Epstein 2011; Maenza 2003
Zygomycetes	Zygomycosis; mucormycosis; opportunistic infection	Decaying vegetables; fruits; soil and animal excreta	Dromer and McGinnis 2003; Epstein 2011; Kwon-Chung and Bennett 1992

predominant organisms were *Staphylococcus aureus, Staphylococcus epidermis, Escherichia coli, Bacillus* spp., and *Proteus mirabilis*. The fungal species were *Aspergillus* spp., *Penicillium* spp., *Mucor* spp., *Candida* spp., and *Verticillium* spp. (Ekhaise, Ighosewe, and Ajakpori 2008).

Rocha et al. (2012), in Venezuela, reported on bioaerosols in a general hospital. They found that the nephrology surgery room, sterilization room, and neonatal room had bacterial and fungal contamination. The densities of microorganisms ranged from 1 to 222 CFU/m^3. Bacterial contamination revealed 14 genera and 8 species of microorganisms. *Staphylococcus* was the most frequent bacteria. *Aspergillus* and *Penicillium* spp. were the most common fungus species among 12 genera and 5 species (Rocha et al. 2012). There have been numerous other references (Augustowska and Dutkiewicz 2006; Nourmoradi et al. 2012; Roberts et al. 2006; Qudiesat et al. 2009). In an article, Fletcher et al. (n.d.) provided information on the importance of bioaerosols in relation to hospital infections, their dispersal, potential microorganisms involved, methods of monitoring bioaerosols, and the control of nosocomial infections using ultraviolet (UV) germicidal irradiation.

There is considerable literature on the health effects of bioaerosols in the indoor environment (Burge 1990; Verhoeff and Burge 1997; Epstein 1997). Mandal and Brandl (2011) pointed out that the presence of human beings performing various activities, such as walking, talking, coughing, sneezing, washing, and toilet flushing, can release airborne biological particles into the air. Furthermore, plants, food material, house dust, pets, textiles, carpets, wood material, and furniture stuffing can occasionally release various fungal spores into the air (Mandal and Brandl 2011). Following flooding, there can be a higher concentration of bioaerosols (Fabian et al. 2005).

Bioaerosols are ubiquitous; consequently, in any indoor working environment where there is moisture there may be the potential for molds, fungi, and other bioaerosols. Several industries have reported on the presence of bioaerosols, often causing illnesses in workers. Examples are agriculture, swine, or pig farming (Radon et al. 2002; Masclaux et al. 2013; Basinas et al. 2013); livestock (Dungan 2010); and greenhouses (Madsen et al. 2013). Bioaerosols are found in hospital environments as well (Fletcher et al. 2006; Drew et al. 2006).

Bioaerosols from composting operations are of great concern. The composting industry in the United States has increased considerably over the past few years due to the ban on landfilling green material. In addition, there is an interest and activity in composting food waste and biosolids (processed sewage sludge), the latter often done in enclosed facilities to reduce the public's exposure to bioaerosols. Workers need to be protected. In the United Kingdom, compost production in 2000–2001 increased from 1 million tonnes to 3.4 million tonnes by 2005–2006 and is expected to grow. The bioaerosols of interest include *Aspergillus fumigatus*, endotoxin, glucans, thermophilic actinomycetes, and mycotoxins.

ASPERGILLUS FUMIGATUS

Aspergillus fumigatus occurs worldwide. It has the ability to degrade cellulose and is found in high numbers (Swan et al. 2003, p. 543). It has been associated with the outdoors (grass, hay, birds' nests and bird droppings, cattle and horse manure, forest litter, wood chips) (Passman 1980) and indoors (refrigerators, bathroom walls, basements, bedding, house dust). It is of greatest importance in composting and is produced in abundance. It is heat tolerant, surviving composting temperatures, and the spores are easily dispersed in the air and are very small so they can easily reach the lung. It is one of the few fungi that can survive human temperatures.

Aspergillus fumigatus grows rapidly at temperatures between 30°C and 52°C and as high as 55°C (Latgé 1999). The fungal spores typically range from 2 to 50 μm. *A. fumigatus* is found in many environments other than composting, and its large presence has a low risk of allergic response for healthy individuals (Drew et al. 2009). However, it is a fungus that can cause allergies and is an opportunistic pathogen that can affect immunocompromised individuals (Swan et al. 2003). It is essential in recycling carbon and nitrogen. It is estimated that all humans will inhale at least several hundred *A. fumigatus* spores per day without harm (Swan et al. 2003). It can also produce mycotoxins, which are low-molecular-weight toxic secondary metabolites that can cause acute or chronic disease in vertebrate animals, with effects ranging from neurotoxicity, to carcinogenicity, and to teratogenicity (Swan et al. 2003). Although we inhale hundreds of spores daily, we are not infected with the organism. Inhalation of conidia by immunocompromised and immunocompetent individuals can have an adverse effect (Latgé 1999).

Endotoxin

An endotoxin is a lipopolysaccharide (LPS) that is part of the cell wall of Gram-negative bacteria. Rylander and Jacobs (1994) indicate that endotoxins are made of complex LPS compounds that consist of polysaccharide chains connected by a core oligosaccharide to a lipid part. Endotoxins are relatively heat stable. They are released into the environment during cell growth and after the cell dies, when the integrity of the cell wall is ruptured (Bradley 1979). They are present in the cavities and intestinal tracts of humans and animals (Swan et al. 2003).

Endotoxin is toxic to humans and animals. Endotoxin can cause fever and short-term illness, with flu-like symptoms, myalgia, and malaise. Inhaled endotoxins increase the activity of macrophages, which leads to a series of inflammatory conditions (Rylander 2002; Swan et al. 2003; Epstein 2011. Rylander (2002) indicates that internalization of endotoxin in macrophages and endothelial cells results in local production of inflammatory cytokines. The result is the migration of inflammatory cells into the lung and the penetration of cytokines into the blood. These lead to inflammation, toxic pneumonia, and systemic symptoms.

Endotoxins present in organic dust have been implicated in toxic pneumonia and organic dust toxic syndrome (Rylander et al. 1989; Rylander and Jacobs 1994; National Institute for Occupational Safety and Health [NIOSH] 2002). Chronic exposure to endotoxin has been linked to work-related symptoms, such as inflammation leading to chronic bronchitis, chronic obstructive pulmonary disease, and reduced lung function (Jacobs et al. 1997).

There have been studies of Dutch compost workers that reported excess acute and subchronic nonimmune inflammation in the upper airways, presumably by exposure to endotoxin (Douwes et al. 2003; Drew et al. 2009; Deacon et al. 2009). Endotoxin was not detected in upwind samples but was consistently detected in on-site samples. Concentrations of endotoxin were consistently higher during composting activities compared to periods of no activity. Downwind measurements showed a pattern of peaks and troughs, suggesting complex dispersal dynamics. Quantities of endotoxin measured on site were consistently below 50 EU/m^3, which the Netherlands suggest as a threshold level for occupational exposure. Cytokine release levels were minimal at concentrations less than 50 EU/m^3 (Deacon 2009). There are no data indicating carcinogenic, mutagenic, or reproductive effects from exposure to endotoxins.

Glucans

(1–3)β-D-glucan is a polyglucose compound in the cell walls of fungi, some bacteria, and plants. It is a potential inflammatory agent that induces nonspecific inflammatory reactions and may also be a respiratory immunomodulatory agent. It may be involved in contributing to the inflammatory response, resulting in respiratory symptoms and adverse lung function effects in response to inhalation of bioaerosols (Swan et al. 2003). It has been found in green waste composting activities and has been proposed as a nonspecific indicator of fungal exposure.

In a study in the Netherlands, an association was found between peak flow variability and (1–3)β-D-glucan levels and house dust among children (Douwes et al. 2000).

Noncellulosic β-glucans are recognized as potent immunological activators. Chen and Seviour (2007) report that they are used in China and Japan in clinics. They further report that the literature suggests β-glucans are effective in treating cancer, a range of microbial infections, hypercholesterolemia, and diabetes and appear to stimulate the immune system (Chen and Seviour 2007). Akramiene et al. (2007) indicate that β-glucans can prevent oncogenesis due to a protective effect against potent genotoxic carcinogens. β-Glucans act on several immune receptors, including dectin-1, complement receptor (CR3), and TLR-2/6, and trigger a group of immune cells, including macrophages, neutrophils, monocytes, natural killer cells, and dendritic cells (Chan, Chan, and Sze 2009). Vetvicka and Vetvickova (2010) tested 16 different glucans to evaluate whether individual glucans were similarly active against each of the tested biological properties or if each glucan effected different reactions. No direct connection between sources and immunological activities were found.

Actinomycetes

Actinomycetes are one of the most diverse groups of filamentous bacteria. They are a subgroup of actinobacteria, which are Gram-positive organisms. They include many familiar and important bacteria, such as *Mycobacterium,* which causes tuberculosis and leprosy. Another bacterium, *Corynebacterium*, can be found in the mucosa and normal skin flora of humans. *Streptomyces* is a source of numerous antibiotics and are the organisms producing the characteristics of freshly turned soil that are essential in the composting process (Burge 2008). *Nocardia asteroids* are actinomycetes that are common in soil and can cause a respiratory infection.

Thermophilic actinomycetes are the most common cause of hypersensitive pneumonitis (e.g., farmer's lung), which results from exposure to hay that has been colonized with thermophilic actinomycetes. They are common in soil and water. The odor associated with fresh soil is attributed to a volatile organic compound, geosmin, produced by actinomycetes.

Actinomycetes are a major component of bioaerosols emitted during composting (Lacey and Dutkiewicz 1994; Taha et al. 2007). Several, such as *Nocardia* spp., *Streptomyces* spp., *Thermoactinomyces vulgaris*, *Thermoactinomyces sacchari*, *Thermomonospora curvata*, and *Thermomonospora* spp. have been identified in composting material (Lacey 1973; Palmisano and Barlaz 1996). Mesophilic actinomycetes grow at temperatures of 20°C to 50°C, whereas thermophilic actinomycetes thrive at 30°C to 60°C with short chains of spherical spores 0.7–1.3 µm in diameter (Taha et al. 2007). Thermophilic actinomycetes such as *Streptomyces* spp., *Thermomonospora* spp., and *Thermoactinomyces vulgaris* can tolerate temperatures in the range of 40°C to 50°C and are found even at 65°C. They prefer moist, highly aerobic, and neutral or slightly alkaline pH conditions (Lacey 1973; Palmisano and Barlaz 1996).

Actinomycetes offer the most promising synthesizers of many industrial and commercially meaningful metabolites (Nawani et al. 2013).

Prolonged inhalation of these bioaerosols is linked to adverse health effects (Douwes et al. 2003), including allergic alveolitis and other respiratory effects (Millner 1982; Lacey and Crook 1988; (Lacey and Dutkiewicz 1994; Herr et al. 2003) in which inflammation of the lung is believed to be caused by glycopeptides and protein allergens caused by spores (e.g., farmer's lung disease) (Edwards 1972.).

Actinomycetes spores are more difficult to aerosolize than fungal spores because of their smaller size (Reponen et al. 1997, 1998).

Mycotoxins

A mycotoxin is a toxic secondary metabolite produced by organisms of the fungi kingdom, commonly known as molds. The term *mycotoxin* is usually reserved for the toxic chemical products produced by fungi that readily colonize crops. One mold species may produce many different mycotoxins, and the same mycotoxins may be produced by several species ("Mycotoxins" n.d.).

Mycotoxins can appear in the food chain as a result of fungal infection of crops, either by being eaten directly by humans or by use as livestock feed.

Mycotoxins greatly resist decomposition or being broken down in digestion, so they remain in the food chain in meat and dairy products. Even temperature treatments, such as cooking and freezing, do not destroy some mycotoxins.

Mycotoxins are small, approximately 0.1 µm. Toxic mold will produce mycotoxins, which are airborne and cause respiratory problems.

Stachybotrys (toxic black mold) is associated with poor air quality, resulting often from water-damaged conditions. Exposure to the mycotoxins present in *Stachybotrys chartarum* or *Stachybotrys atra* can have a wide range of effects. Depending on the length of exposure and volume of spores inhaled or ingested, symptoms can manifest as chronic fatigue or headaches; fever; irritation to the eyes, mucous membranes of the mouth, nose, and throat; sneezing; rashes; and chronic coughing. In severe cases of exposure or cases exacerbated by allergic reaction, symptoms can be extreme, including nausea, vomiting, and bleeding in the lungs and nose ("*Stachybotrys*" n.d.). Recently, there has been heightened concern regarding exposure to a specific type of mold commonly referred to as black or toxic mold. However, currently there is no conclusive scientific evidence linking the inhalation of black mold spores or any type of mold in the indoor environment to any illness other than allergy symptoms. The term *toxic* is inaccurate. Some indoor molds are capable of producing extremely potent toxins (mycotoxins) that are lipid soluble and readily absorbed by the intestinal lining, airways, and skin. These agents, usually contained in the fungal spores, have toxic effects, ranging from short-term irritation to immunosuppression and cancer. Among the indoor mycotoxin-producing species of molds are *Fusarium*, *Trichoderma*, and, although less commonly isolated, *Stachybotrys atra* (also known as *S. chartarum*, black mold). There is almost a complete lack of information on specific human responses to well-defined exposures to mold contaminants. There is currently no proven method to measure the type or amount of mold to which a person is exposed since common symptoms associated with mold exposure are nonspecific. Molds are present everywhere in the environment, particularly where there is moisture (e.g., under sinks), and responses to exposure vary greatly among individuals. Molds often cause allergic symptoms. The concentration at which molds have an impact on human health is unknown (Robbins et al. 2000).

REFERENCES

Airborne Microorganisms. n.d. http://archive.bio.ed.ac.uk/jdeacon/microbes/#choice

Akramiene G, Kondrotas A, Didziapetriene J, and Kevelaitis E. 2007. Effects of beta-glucans on the immune system. *Medicina (Kaunas)* 43(8): 597–606.

American Conference of Governmental Industrial Hygienists (ACGIH). 1999. *Bioaerosols: Assessment and Control*. Cincinnati, OH: ACGIH.

Arnow PM, Anderson RL, Mainous PD, and Smith EJ. 1978. Pulmonary aspergillosis during hospital renovation. *Am Rev Respir Dis* 118: 49.

Augustowska M and Dutkiewicz J. 2006. Variability of airborne microflora in a hospital ward within a period of one year. *Ann Agric Environ Med* 13: 99–108.

Basinas I, Schlunssen V, Takai H, Heederik D, Omland D, Wouters IM, Sigsgaard T, and Kromhout H. 2013. Exposure to inhalable dust and endotoxin among Danish pig farmers affected by work tasks and stable characteristics. *Ann Occup Hyg* 57: 1005–1019.

Bodey GP and Vartibvarian S. 1989. Aspergillosis. *Eur J Clin Microbiol Infect Dis* 8: 413–437.

Bradley SG. 1979. Cellular and molecular mechanisms of action of bacterial endotoxins. *Am Rev Microbiol* 33: 67–94.

Burge HA. 1990. Bioaerosols: prevalence and health effects in the indoor environment. *J Allergy Clin Immunol* 86: 687.

Burge HA. 2008. Actinomycetes. *Environ Rep* 6: 1–4.

Burton JR, Zachery JB, Bessin R, Rathbun HK, Greenough WB, Sterioff S, Wright JR, Slavin RE, and Williams GM. 1972. Aspergillosis in four renal transplant recipients. *Ann Intern Med* 77: 383–388.
Chan OC, Chan WK, and Sze DM. 2009. The effect of beta-glucan on human immune and cancer cells. *J Hematol Oncol* 2: 25.
Chen J and Seviour R. 2007. Medicinal importance of fungal beta-(1-6)-glucans. *Mycol Res* 111(Pt 6): 635–652.
Cox CS, and Wathes CM, ed. 1995. *Bioaerosol Handbook*. Boca Raton, FL: CRC Press.
Dagnani MC, Kiwan EN, and Anaissie EJ. 2003. Hyalohyphomycosis. In *Clinical Mycology*, ed. Anaissie EJ, McGinnis MR, Pfaller MA. Philadelphia: Churchill Livingstone, 319.
Dagnani MC, Solomkin JS, and Anaissie EJ. 2003. *Candida*. In *Clinical Mycology*, ed. Anaissie EJ, McGinnis MR, Pfaller MA. Philadelphia: Churchill Livingstone.
Deacon J. 2009. *The Microbial World: Institute of Cell and Molecular Biology*. Edinburgh: University of Edinburgh.
Deacon L, Parkhurst L, Liu J, Gewq GH, Hayes ET, Jackson S, Longhurst J, Longhurst P, Pollard S, and Tyrrel S. 2009. Endotoxin emissions from commercial composting activities. *Environ Health* 8: 51–59.
Domsch KH, Gams W, Anderson TH. 1980. *Compendium of Soil Fungi*. London: Academic Press.
Douwes ZA, Doekes G, van der Zee SC, Wouters I, Boezen MH, and Runekreef B. 2000. (1-3)-beta-D-glucan and endotoxin in house dust and peak low variability in children. *Am J Respir Crit Care Med* 162: 1348–1354.
Douwes J, Thorne P, Pearce N, and Hendrick D. 2003. Bioaerosols health effects and exposure assessment: progress and prospects. *Ann Occup Hyg* 47: 187–200.
Drew GHD, Pankhurst LJ, Pollard SJT, and Tyrrel S. 2009. *Guidance on the Evaluation of Bioaerosol Risk Assessments for Composting Facilities*. Bedford, UK: Cranfield University.
Drew GHD, Tamer A, Taha MPM, Smith R, Longhurst PJ, Kinnersley R, and Pollard SJT. 2006. Dispersion of bioaerosols from composting facilities. Proceedings of Waste 2006 Conference on Integrated Waste Management and Pollution Control, Stratford-upon-Avon, UK.
Dromer F and McGinnis MR. 2003. Zygomycosis. In *Clinical Mycology*, ed. Anaissie EJ, McGinnis MR, Pfaller MA. New York: Churchill Livingston.
Dungan RS. 2010. Fate and transport of bioaerosols associated with livestock operations and manures. *J Anim Sci* 88: 3693–3706.
Edwards JH. 1972. The isolation of antigen associated with farmer's lung. *Clin Exp Immunol* 11: 341–355.
Ekhaise FO, Ighosewe OU, and Ajakpori OD. 2008. Hospital indoor airborne microflora in private and government owned hospitals in Benin City, Nigeria. *World J Med Sci* 3: 34–38.
Epstein E. 1997. *The Science of Composting*. Lancaster, PA: Technomic.
Epstein E. 2011. *Industrial Composting*. Boca Raton, FL: CRC Press.
Fabian MP, Miller SL, Reponen T, and Hernandez MT. 2005. Ambient bioaerosol indices for indoor air quality assessments of flood reclamation. *Aerosol Sci* 36: 763–783.
Fletcher LA, Noakes CJ, Beggs CB, and Sleigh PA. 2006. The importance of bioaerosols in hospital infections and the potential for control using germicidal ultraviolet irradiation. http://www.efm.leeds.ac.uk/CIVE/aerobiology/PDFs/uv-and-airborne-hospital-infection-fletcher.pdf
Herr CEW, zur Nieden A, Jankofsky V, Stilianakis NJ, Boedeker RH, and Eikmann TF. 2003. Effects of bioaerosol polluted outdoor air on airways of residents: a cross sectional study. *Occup Environ Med* 60: 336–342.

Hopkins CC, Webber DJ, and Rubin RH. 1989. Invasive *Aspergillus* infection: possible nonward source within the hospital environment. *J Hosp Infect* 13: 19–25.
Jacobs RR, Heedrick D, Douwes J, and Zahringer U. 1997. Endotoxin structure. *Int J Occup. Environ Health* 3: S6–S7.
Kwon-Chung KJ and Bennett JE. 1992. *Medical Mycology*. Philadelphia, PA: Lea and Febiger.
Lacey J. 1973. Actinomycetes in soils, composts and fodder. In *Actinomycetales: Characteristics and Practical Importance*. London: Academic Press, 231.
Lacey J, and Crook B. 1988. Fungal and actinomycetes spores as pollutants of the workplace and occupational allergens. *Ann Occup Hyg* 32: 515–533.
Lacey J, and Dutkiewicz J. 1994. Bioaerosols and occupational lung disease. *J Aerosol Sci* 25: 1371–1404.
Latgé JP. 1999. *Aspergillus fumigatus* and aspergillosis. *Clin Microbiol Rev* 12: 210–350.
Levetin E. n.d. Airborne fungi in indoor environments. http://pollen.utulsa.edu/Spores/Indoor.htm
Madsen AM, Tendal K, and Frederiksen MW. 2014. Attempts to reduce exposure to fungi, β-glucan, bacteria, endotoxin and dust in vegetable greenhouses and a packaging unit. *Sci Total Environ* 468/469: 1112–1121.
Mandal J and Brandl, H. 2011. Bioaerosol in indoor environment—a review with special reference to residential and occupational locations. *Open Environ Biol Monitor J* 4: 83–96.
Masclaux FC, Sakwinska O, Charriere N, Semaani E, and Oppliger A. 2013. Concentration of airborne *Staphylococcus aureus* (MSRA and MSSA), total bacteria, and endotoxins in pig farms. *Ann Occup Hyg* 57: 550–557.
Maxmen A. 2011. Asthma: breathing new life into research. *Nature* 479: S20–S21.
McGinnis MR and Pfaller MA, eds. 2009. *Clinical Mycology*. New York: Churchill Livingston.
Millner PD. 1982. Thermophilic and thermotolerant actinomycetes in sewage sludge compost. Symposium: non-antibiotic producing actinomycetes. *Dev Indust Microbiol* 23: 61–78.
Mullins J, Harvey R, and Seaton A. 1976. Sources and incidence of airborne *Aspergillus fumigatus* (Fres.). *Clin Allergy* 6: 209–217.
Mycotoxins. n.d. http://en.wikipedia.org/wiki/Mycotoxin.
National Institute for Occupational Safety and Health (NIOSH). 2002. *Request for Assistance in Preventing Organic Dust Toxic Syndrome*. Report no. 94-102. Cincinnati, OH, National Institute for Occupational Safety and Health.
Nawani N, Aigle B, Mandal A, Bodas M, Ghotnrl S, and Prakash D. 2013. Actinomycetes role in biotechnology and medicine. *BioMed Res Int* article ID687190. http://dx.doi.org./10.1155/2013/687190
Nobel WC, and Clayton YM. 1963. Fungi in the air of hospital wards. *J Gen Microbiol* 32: 397–402.
Nourmoradi H, Amin MM, Hatamzadeh M, and Nikaeen M. 2012. Evaluation of bio-aerosols concentrations in the different wards of three educational hospitals in Iran. *Int J Env Health Eng* 1: 47.
Palmisano AC, and Barlaz MA, eds. 1996. *Microbiology of Solid Waste*. Boca Raton, FL: CRC Press.
Passman FJ. 1980. *Monitoring of* Aspergillus fumigatus *Associated with Municipal Sewage Sludge Composting Operations in the State of Maine*. Portland, ME: Portland Water District.
Pfaller MA, Diekerma DJ, and Merz WG.. 2003. Infections caused by non-*Candida*, non-*Cryptococcus* yeasts. In: *Clinical Mycology*, ed. Anaissie EJ, McGinnis MR, Pfaller MA. Philadelphia: Churchill Livingston.
Polymenakou PN. 2012. Atmosphere: a source of pathogenic or beneficial microbes? *Atmosphere* 3: 87–102.

Prasad M, van der Werf P, and Brickman A. 2004. Bioaerosols and composting—a literature evaluation. Composting Association of Ireland. http://www.cre.ie/docs/cre_bioaerosol_aug2004.pdf (accessed February 18, 2011).

Qudiesat K, Abu-Elteen K, Elkarmi A, Hamad M, and Abussaud M. 2009. Assessment of airborne pathogens in healthcare settings. *African J Microbiol Res* 3: 066–076.

Radon K, Monso E, Weber C, Danuser B, Iverson M, Opravil U, Donham K, Hartung J, Pedersen S, Garz S, Blainey D, Rabe U, and Nowak D. 2002. Prevalence and risk factors for airway diseases in farmers—summary of results of the European Farmer's Project. *Ann Agric Environ Med* 9: 207–213.

Reponen TA, Gazenko SV, Grinshpun SA, Willeke K, and Cole EC. 1998. Characteristics of airborne actinomycetes spores. *Appl Environ Microbiol* 64(10): 3807–3812.

Reponen TA, Willeke K, Ulevicius V, Grinshpun SA, and Donnelly J. 1997. Techniques for dispersion of microorganisms into air. *Aerosol Sci Technol* 27: 405–421.

Rhodes JC. 2006. *Aspergillus fumigatus*: growth and virulence. *Med Mycol* 44: 577–581.

Robbins CA, Swenson LJ, Nealley ML, Kelman BJ, and Gots RE. 2000. Health effects of mycotoxins in indoor air: a critical review. *Appl Occup Environ Hyg* 15: 773–784.

Roberts A, Hathway A, Fletcher LA, Beggs CB, Elliott MW, and Sleigh PA. 2006. Bioaerosol production on a respiratory ward. *Indoor Built Env* 15: 35–40.

Rocha CA, Baez NA, Villarroel EV, and Quintero GM. 2012. Study of bioaerosols in surgical theaters and intensive care units from a public general hospital. *J Biosci Med* 2(3). doi: 10.5780/jbm2012.26

Rodenhuis S, Beaumont F, Kauffman HF, and Sluiter HJ. 1984. Invasive pulmonary aspergillosis in a non-immunosuppressed patient: successful management with systemic amphotericin and flucytosine and inhaled amphotericin. *Thorax* 39: 78–79.

Rylander R. 2002. Endotoxin in the environment—exposure and effects. *J Endotoxin Res* 8: 241–252.

Rylander R, Bake B, Fisher JJ, and Helander IM. 1989. Pulmonary function and symptoms after inhalation of endotoxin. *Am Rev Respir Dis* 140: 981–986.

Rylander R and Jacobs RR. 1994. *Organic Dust Exposure, Effects, and Prevention*. Boca Raton, FL: Lewis.

Stachybotrys. n.d. http://en.wikipedia.org/wiki/Stachybotrys

Summerbell RC, Krajden S, and Kane J. 1989. Potted plants in hospitals as reservoirs of pathogenic fungi. *Mycopathologia* 106: 13–22.

Swan JRM, Crook B, and Gilbert EJ. 2002. Microbial emissions from composting sites. *Environ Sci Technol* 18: 73–101.

Swan JRM, Kelsey A, Crook B, and Gilbert EJ. 2003. *Bioaerosol Components and Hazards to Human Health. A Critical Review of Published Data*. Report no. 130. Sheffield and Northhamptonshire, UK: Health and Safety Laboratory and The Composting Association.

Taha MPM, Drew GH, Tamer Vestlund A, Aldred D, Longhurst PJ, and Pollard SJT. 2007. Enumerating actinomycetes in compost bioaerosols at source—use of soil compost agar to address plate "masking." *Atmos Environ* 41(22): 4759–4765.

Verhoeff AP and Burge HA. 1997. Health risks assessment of fungi in home environments. *Ann Allergy Asthma Immunol* 78: 544–554.

Vetvicka V and Vetvickova J. 2010. β1,3-Glucan: silver bullet or hot air? *Open Glycosci* 3: 1–6.

Viviani MA, Tortorano AM, and Ajello L. 2003. Cryptococcus. In *Clinical Mycology*, ed. Anaissie EJ, McGinnis MR, Pfaller MA. Philadelphia: Churchill Livingston.

Walsh TJ, Groll A, Hiemenz J, Fleming R, Roilides E, and Anaissie E. 2004. Infections due to emerging and uncommon medically important fungal pathogens. *Clin Microbiol Infect* 10: 48–76.

World Health Organization (WHO). 2013. Asthma. http://www.who.int/mediacentre/factsheets/fs307/en/index.html

9 Pathogens in Animal Waste and Manures

INTRODUCTION

In 2009, the Water Environment Federation published a book, *Manure Pathogens*, edited by Dr. Dwight D. Bowman. This chapter updates the information presented in that book (Bowman 2009). The pathways of sources and transmission of pathogens from animals to humans are shown in Figure 9.1. If manure is not treated to destroy pathogens, then they can survive in the environment for a long time. Table 9.1 illustrates the survival times of several pathogens in the environment.

More than 85% of the world's fecal wastes are from domestic animals, such as poultry, cattle, sheep, and pigs. These animals harbor zoonotic pathogens. Diseases passed from animals to humans are called zoonoses. These pathogens can enter the environment and infect humans. Primarily, this environmental contamination is through runoff. However, as indicated previously in this book, unless the wastes are incorporated into the ground, there could be leaching and movement to drinking water resources. Little information exists on health effects associated with exposure to this potential hazard to human health.

Olson (2001) reported that six prevalent enteric pathogens were found in cattle, pigs, and poultry (Table 9.2). Thus, manures from these animals could harbor these organisms and, if in contact with humans, cause serious illnesses.

One of the major criteria for potential infection to humans from pathogens is their infective dose. This refers to the number of pathogens required to infect a host or the number of organisms required to cause a disease. With humans, this is highly variable. Factors such as age, individual immune conditions, route of infection, and route of transmission (inhalation, dermal absorption, ingestion) are some of the important ones. The US Food and Drug Administration, in their *Bad Bug Book* (2013), indicates that the infective dose values should be viewed with caution for the following reasons:

- Often, they are extrapolated from epidemiological investigations.
- They were obtained by human feeding studies of healthy, young adult volunteers.
- They are best estimates based on a limited database from outbreaks.
- They are worst-case estimates.
- There are several variables that cannot be directly used to assess risks.

The following are some of the variables:
Variables with regard to microorganisms:
- Variability of gene expression of multiple pathogenic mechanisms
- Potential for damage or stress of the microorganism

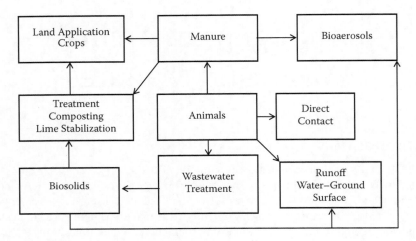

FIGURE 9.1 Pathways of sources and transmission of pathogens from animals to humans.

TABLE 9.1
Survival of Several Animal Pathogens in the Environment

Pathogens	Soil	Cattle Manure	Grass
Listeria	2 weeks to 4 years		128 days
Escherichia coli	1 month to 1 year	10 days to 3 months	99 days
Cryptosporidium	1 month to 1 year	4 weeks to 1 year	30 days
Salmonella	2 weeks to 1 month	4 weeks to 6 months	63 days
Campylobacter	1 month	1 week to 8 weeks	

Source: Wisconsin—Farm Safety Fact Sheet, University of Wisconsin, Madison.

TABLE 9.2
Prevalent Enteric Pathogens

Pathogen	Cattle (%)	Pigs (%)	Poultry (%)
Salmonella spp.	0–13	0–38	10–100
Escherichia coli 0157:H7	16	0.4	1.3
Campylobacter jejuni	1	2	100
Yersinia enterocolitica	<1	18	0
Giardia lamblia	10–100	1–20	0
Cryptosporidium spp.	1–100	0–10	0

- Interaction of organism with food solvent and environment
- The organism's susceptibility to pH
- The organism's immunological uniqueness
- Interactions with other organisms

Variables with regard to the host:
- Age
- General health
- Pregnancy
- Medication: over the counter or prescription
- Metabolic disorders
- Alcoholism, cirrhosis (progressive disease of the liver), hemochromatosis (a disease of older men with excessive iron intake)
- Malignancy
- Amount of food consumed
- Gastric acidity variation
- Genetic disturbances
- Nutritional status
- Immune competence
- Surgical history
- Occupation

The following are some infective dose data:

- *Shigella* and *Giardia*: approximately 10 cells to start an infection (Leggett, Cornwallis, and West 2012)
- *Vibrio cholera* and *Staphylococcus*: 10^3 to 10^8 cells to start an infection (Leggett, Cornwallis, and West 2012)
- *Salmonella* and strains of *Escherichia coli*: 10^5 or more organisms (Kothary and Babu 2001)
- Toxigenic *Vibrio cholerae* (01 and 0139 serotypes): 10^4 organisms (Kothary and Babu 2001)
- *Cryptosporidium jejuni*: Infective dose as low as 500 organisms (Kothary and Babu 2001)
- *Cryptosporidium parvum*: 10 oocysts (Kothary and Babu 2001)
- *Entamoeba coli*: One cyst (Kothary and Babu 2001)
- *Escherichia coli*: Approximately 10 cells (Schmid-Hempel and Frank 2007)
- *Variola* (smallpox virus): One virus particle (Nicas et al. 2004)

PATHOGENS AND DISEASES TRANSMITTED TO HUMANS

The pathogens transmitted from animals to humans can be bacteria, viruses, and parasites. More than 150 pathogens can cause zoonotic infections (Jensen 2011). Some of the most important ones are discussed next.

BACTERIA

- *Escherichia coli*: Pathogenic strains cause enteritis, peritonitis, or cystitis.
- *Salmonella* spp.: Some species are pathogenic in animals and humans. They can cause food poisoning.
- *Leptospira* spp.: These bacteria can cause leptospirosis, a fever that is the world's most common disease transmitted to people from animals. The infection is commonly transmitted to humans by allowing water that has been contaminated by animal urine to come in contact with unhealed breaks in the skin, the eyes, or the mucous membranes.
- *Aeromonas*: Some species are pathogenic. Two major diseases associated with *Aeromonas* are gastroenteritis and wound infections, with or without bacteremia.
- *Listeria monocytogenes*: One species can cause meningitis; encephalitis; septicemia; endocarditis; abortion; abscesses; and lesions. It is often fatal.
- *Yersinia enterocolitica*: Contamination from feces occurs. Symptoms may include watery or bloody diarrhea and fever.
- *Vibrio* spp.: *Vibrio cholerae* is the cause of cholera; *Vibrio sputorum* causes diseases of the mouth.
- *Campylobacter* spp.: These species can result in campylobacteriosis, a common cause of food poisoning.
- Hydrophilia (pathogenic strains): *Aeromonas hydrophila* can cause gastroenteritis.

VIRUSES

- Hepatitis E (swine): This has resemblance to the human hepatitis E virus and its ability to infect nonhuman primates.
- Rotaviruses: Rotavirus is a virus that causes gastroenteritis (inflammation of the stomach and intestines). It can cause diarrhea, vomiting, fever, and abdominal pain. Rotavirus is the most common cause of severe gastroenteritis in infants and young children.
- Adenoviruses: These are associated with respiratory infections of children, conjunctivitis, and pneumonia.
- Caliciviruses (certain strains): Influenza viruses (orthomyxoviruses), Norwalk virus, and other caliciviruses produce infections that cause acute diarrhea and vomiting (gastroenteritis), abdominal cramps, myalgia, malaise, headache, nausea, and low-grade fever.

PARASITES (PROTOZOANS)

- *Cryptosporidium parvum*: This is one of several protozoa species that cause cryptosporidiosis, a parasitic disease of the human intestinal tract.

- *Giardia lamblia*: This is a parasite of the small intestinal tract. It interferes with the absorption of fats and produces flatulence, steatorrhea, and acute discomfort.
- *Balantidium coli*: A parasitic ciliate, it is a cause of colitis similar to amebic dysentery.

It has been reported that potentially harmful pathogens, including verocytoxigenic *Escherichia coli* (VTEC) are shed in animal manures, and there is growing concern in many countries about the number of sporadic and outbreak cases of VTEC attributable to direct contact with fecal material either as a result of handling contaminated soil in fields or ingestion of produce grown in contaminated manures or slurries. VTEC has been detected in the feces of ruminant and nonruminant farmed animals, wild animals, domestic pets, and birds. This pathogen appears to be well adapted to survive in animal wastes and can persist for extended periods, ranging from several weeks to many months. Because of this persistence, these animal manure materials are important as potential vehicles for transmission within farm animals, fresh food, and the wider environment.

The application of raw manure to crops to be eaten raw needs to be avoided. Several treatments are available to disinfect manure. The most economical ones for farmers or other producers of manure are composting and lime stabilization. Depending on location (i.e., with respect to residences or others affected by odors or noise), composting may be done in an enclosed facility or outdoors. The common composting methods are windrowing, passive aeration, or aerated static piles. Windrow composting is typically conducted outdoors and unless using fabric covers can result in severe odors (see Figure 9.1).

Passive aeration is the least expensive, but there is little experience and confirmation regarding pathogen disinfection. The aerated static pile, originally developed by the US Department of Agriculture, has been primarily used with sewage sludge or biosolids. The data on disinfection is extensive and well documented.

In the past, farmers piled manure for several months or years, and since there is a relationship between length of time and disinfection, considerable die-off of pathogens occurred. However, this process required considerable space and often offended nearby residences. Unlike municipalities that have considerable resources, farmers have limited finances.

The survival of several pathogens found in animal fecal wastes are shown in Table 9.3. In water, the optimum temperature was 5°C for all the organisms reported. In the soil, cold temperatures also enhanced survival. The surprise was the survival time for the compost. Apparently, this compost did not reach 55°C, as required by both Canadian and US regulations for sewage sludge or biosolids. As expected, survival was low on dry surfaces. It was not specified if the dry surfaces were in the winter or summer.

TABLE 9.3
Survival of Pathogens Found in Animal Fecal Wastes

		Duration of Survival					
Material	Temperature	Giardia	Cryptosporidium	Salmonella	Campylobacter	Yersinia enterolitica	Escherichia coli O157:H7
Water	Frozen	>1 day	>1 year	>6 months	2–8 weeks	>1 year	>300 days
	5°C	11 weeks	>1 year	>6 months	12 days	>1 year	>300 days
	30°C	2 weeks	10 weeks	>6 months	4 days	10 days	84 days
Soil	Frozen	<1 day	>1 year	>12 weeks	2–8 weeks	>1 year	>300 days
	5°C	7 weeks	8 weeks	12–28 weeks	2 weeks	>1 year	100 days
	30°C	2 weeks	4 weeks	4 weeks	1 week	10 days	2 days
Cattle feces	Frozen	<1 day	>1 year	>6 months	2–8 weeks	>1 year	>100 days
	5°C	1 week	8 week	12–28 weeks	1–3 weeks	30–100 days	>100 days
	30°C	1 week	4 week	4 weeks	1 week	10–30 days	10 days
Slurry		1 year	>1 year	13–75 days	>112 days	12–28 days	10–100 days
Compost		2 week	4 week	7–14 days	7 days	7 days	7 days
Dry surfaces		1 day	1 day	1–7 days	1 day	1 day	1 day

Source: Olson ME. 2001. Human and animal pathogens in manure. Paper presented at Livestock Options for the Future, National Conference, Winnipeg, Canada. June 2001.

The Ontario Ministry of Agriculture and Food stated:

Using proper composting procedures can drastically reduce the number of pathogens in manure. One of the parameters identified in many composting procedures is to maintain temperature above 55°C for at least 3 days when using aerated or in-vessel systems. In windrow systems, the core of the windrow may reach these temperatures, but surface zones and near the base of the windrow will have lower temperatures. Turning or mixing the windrow will introduce oxygen to the windrow and quickly increase temperatures in the earlier stages of composting. Turning with equipment that moves material from the surface to the core of the windrow will expose more materials to higher temperatures. Repeated turnings are necessary to ensure all materials are exposed to at least 3 consecutive days of high temperature.

It is generally recommended that windrows maintain a core temperature of 55°C for 15 days with at least 5 turnings. Due to the need for proper mixing and consistent high temperatures, pathogen reduction in windrow composting has sometimes been found to be less consistent than when using well-managed, aerated static pile or in-vessel systems.

Some organisms are more difficult to kill than others and temperatures must be regularly monitored to ensure that appropriate temperatures are maintained. Proper moisture levels (50%–60%), optimum C:N ratios (25:1–30:1) and appropriate aeration or turnings are all factors to ensure the conditions are optimum to reach and maintain the necessary temperatures. Records should be kept for temperatures, compost conditions, C:N ratio, moisture, and the date/time of turnings. (Martin 2005)

Campylobacter was found to survive at temperatures of raw and cooked poultry samples, although its optimum temperature ranged between 37°C and 42°C. A study by Haddad et al. (2009) found that long-term survival of *Campylobacter jejuni* at low temperatures is dependent on polynucleotide phosphorylase activity. It is therefore possible that other pathogens can have characteristics that allow them to survive under nonoptimal conditions in the environment.

In a review, Guan and Holley (2003) indicated that survival of pathogens in soil, manure, and water indicated significant variability in resistance to environmental challenges. Generally, pathogens survive longer in the environment at cooler temperatures, but differences were reported for both liquids and solid manure. Temperatures ranging from 25°C to 90°C destroyed pathogens.

In the developed countries, the major concern is contamination of the water supplies. Effective low-cost treatment such as composting would significantly destroy animal pathogens. Incorporating animal waste to a depth of at least 6 inches (plow zone) will greatly reduce contamination by runoff.

In developing countries, especially in small villages, animal waste could be composted or, if untreated, should not be used to grow food chain crops. These crops could be nuts, cotton, fruit trees, and others. If it is needed or desired to use untreated animal manures on food chain crops, it would be best to apply them to fields that would not be used for weeks or even a year.

SUMMARY

Manure Pathogens (Bowman 2009) is an excellent book on the subject of animal manures. Animal wastes can harbor pathogens that can harm humans and cause diseases and illnesses. Diseases passed from animals to humans are called zoonoses. More than 85% of the world's fecal wastes are from domestic animals, such as poultry, cattle, sheep, and pigs.

The largest waterborne disease outbreak in US history occurred in 1993 in Milwaukee, Wisconsin, when over 400,000 people became ill with diarrhea when the parasite *Cryptosporidium* was found in the city's drinking water supply (Brunkard et al. 2011). This was thought to be the result of manure shed by cattle that resulted in runoff that contaminated the water supply. Another incident attributed to animal waste contamination occurred on May 11, 2000, when many people of a community of about 5,000 people began to simultaneously experience bloody diarrhea, gastrointestinal infections, and other symptoms of *E. coli* O157:H7 and *Cryptosporidium* infection. Two thousand persons became ill, and seven died (Holme, 2003; Clark et al. 2003).

The survival time of pathogens in animal wastes can extend for a year of more depending on the temperature of the waste, the type of waste, and the organism. The pathogens transmitted from animals to humans can be bacteria, viruses, and parasites. More than 150 pathogens can cause zoonotic infections. Treatments that raise the temperature of the waste to over 55°C are effective in destroying animal pathogens.

REFERENCES

Bowman DD. 2009. *Manure Pathogens*. Alexandria, VA: WEF Press.

Brunkard JM, Ailes E, Roberts VA, Hill V, Hilborn ED, Craun GF, Rajasingham A, Kahler A, Garrison L, Hicks L, Carpenter J, Wade TJ, Beach MJ, and Yoder JS. 2011. *Surveillance for Waterborne Disease Outbreaks Associated with Drinking Water—United States, 2007–2008*. Atlanta, GA: CDC.

Clark CG, Rafiq Ahmed LP, Woodward DL, Melito PL, Rodgers FG, Jamieson F, Ciebin B, Li A, and Ellis A. 2003. Characterization of waterborne outbreak–associated *Campylobacter jejuni*, Walkerton, Ontario. *Emerg Infect Dis* 9: 1232–1241.

Guan TY and Holley RA. 2003. Pathogen survival in swine manure environments and transmission of human enteric illnesses—a review. *J Environ Qual* 32: 383–392.

Haddad N, Burns CM, Bolla JM, Pre'vost H, Fe'de'right M, Drider D, and Cappelier JM. 2009. Long-term survival of *Campylobacter jejuni* at low temperatures is dependent on polynucleotide phosphorylase activity. *Appl Env Microbiol* 75: 7310–7318.

Holme R. 2003. Drinking water contamination in Walkerton, Ontario: positive identification. *Water Sci Technol* 47: 1–6.

Jensen J. 2011. Foodborne Disease in the United States. http://vbs.psu.edu/extension/resources/pdf/presentations/pathogens.pdf

Kothary MH and Babu US. 2001. Infective dose of foodborne pathogens in volunteers: a review. *J Food Safety* 21: 149–268.

Leggett HC, Cornwallis CK, and West SA. 2012. Mechanisms of pathogenesis, infective dose and virulence in human parasites. *PLoS Pathog* 8(2): e1002512. doi:10.1371/journal.ppat.1002512

Martin H. 2005. Agricultural Composting Basics. http://www.omafra.gov.ca/english/engineer/facts/05-021.pdf

Nicas M, Hubbard AE, Jones RM, and Reingold AL. 2004. The infectious dose of *Variola* (small pox) virus. *Appl Biosafety* 9: 118–127.

Olson ME. 2001. Human and animal pathogens in manure. Paper presented at Livestock Options for the Future, National Conference, Winnipeg, Canada. June 2001.

Schmid-Hempel P and Frank SA. 2007. Pathogenesis, virulence and infective dose. *PLoS Pathog* 3: e147.

US Food and Drug Administration. 2013. *Bad Bug Book. Handbook of Foodborne Microorganisms and Natural Toxins*. 2nd ed. http://www.fda.gov/food/foodborneillnesscontaminants/causesofillnessbadbugbook/default.htm

Wisconsin Farm Safety Fact Sheet. Madison, WI: University of Wisconsin.

10 Pathogens in Food and Water

INTRODUCTION

This chapter includes both foodborne and waterborne pathogens together because of their relationship. These two need to be managed together to be effective in controlling diseases and human exposure. Food produced can be exposed to pathogens through the soil, water, and air. The soil can harbor pathogens, which can be transmitted to roots and to plants. Some of these pathogens are introduced to the soil through the application of manure, contaminated water, or aerosols.

Water is a key ingredient of foodstuffs, and contamination can occur from various sources. Irrigated contaminated water resulting from both domestic and wildlife excreta can contaminate crops. The World Health Organization (WHO) indicates that human exposure to pathogens from food and water as a result of wastewater reuse in the irrigation of vegetables and fruits resulted in disease outbreaks (WHO 2003). WHO suggests that the use of a common approach for characterization of microbial hazards in food and water will result in greater understanding, which can effectively reduce disease risk and improve public health.

Education and hygiene can significantly result in reducing foodborne and waterborne diseases. This primarily will be effective in developed countries. The population of developed countries tends to have a higher degree of education. Both parents and children are taught the importance of washing hands after using the toilet and prior to eating. Unfortunately, this is not the case in developing countries.

In developing countries, access to infrastructure, flush toilets, clean water, and other sources related to hygiene are not available. The lack of availability of these facilities can result in fecal-to-oral contamination.

There are numerous factors that affect the potential for a pathogen to cause a disease (WHO, 2003). These can include

- The pathogen's properties and characteristics (e.g., ability to survive and multiply in food and water)
- The pathogen's phenotypic and genetic characteristics
- Host specificity: tolerance is generally lower for the young and elderly
- Infection mode
- Antimicrobial resistance
- Environmental conditions: heat, drought

A pathogen's properties and characteristics have been discussed in other chapters. Survival and persistence rates vary with the organism and type of waste and its treatment.

Host-related factors are important and can influence the likelihood of infection. These can affect the probability and severity of disease (WHO 2003). These can include

- Age: Children and the elderly are more susceptible and may have greater difficulty fighting diseases.
- The general health of the individual: In many cases, this can be related to nutrition and the availability of medical care and its implementation.
- The immune status of the individual: Today, in developed countries, immunization can be significant.
- Exposure.
- Nutritional status.
- Demographic conditions: Poverty, social, cultural, and religious impacts.

An *antimicrobial* is defined as an agent or agents that kill organisms or inhibit their growth. These agents could be natural or man-made. Natural antimicrobial agents found in soil and compost inhibit waste-related pathogens. One organism well known for resistance to antibiotics is methicillin-resistant *Staphylococcus aureus* (MRSA) bacteria. There is evidence of a connection of antimicrobial-resistant human infections to foodborne pathogens of animal origin (Swartz 2002). *Salmonella, Campylobacter,* and *Escherichia coli* O157:H7 have been shown to be antimicrobial resistant (Swartz 2002).

Climate change and other environmental conditions, such as surface heat, ultraviolet light, or dryness, can effectively reduce or eliminate pathogens. These conditions can be significant in destroying pathogens in agriculture and food production.

FOODBORNE DISEASES

The foodborne diseases listed under this topic are the most common in the United States and developing countries. Foodborne diseases usually manifest in gastrointestinal problems such as diarrhea and vomiting with the potential for fever. Symptoms could be caused by toxins as a result of the growth of the pathogen in foods or by gastrointestinal infection (Bates, Hoffman, and Morris 2011). According to the University of Florida, symptoms from preformed toxins can generally have a fairly rapid onset. Often, the effect does not show up for several hours, or it takes a day or two for symptoms to appear. Diseases caused by bacteria and viruses tend to show up more rapidly, whereas those as a result of protozoa are slower.

Some of the more common foodborne pathogens, according to the Iowa State University Food Safety Project, the US Food and Drug Administration (FDA) (2014), and Epstein (2011), are the following:

- *Bacillus cereus*: It is responsible for a minority of foodborne illnesses (2%–5%), causing severe nausea, abdominal discomfort, vomiting, and diarrhea. *Bacillus* foodborne illnesses occur due to survival of the bacterial endospores when food is improperly cooked.

- *Campylobacter jejuni*: This bacterium is the most common cause of diarrhea in the United States. The highest rate is found in children under the age of 1 year. Unborn babies and infants are more susceptible on exposure to this organism. The sources are raw milk; contaminated water; raw and undercooked meat, poultry, or shellfish; and eggs. The incubation period is generally 2 to 5 days after eating contaminated food and lasts for 7 to 10 days.
- *Clostridium botulinum*: This bacterium produces a toxin that causes botulism, resulting in muscular paralysis. Honey can contain *Clostridium botulinum* spores. Sources can be home-canned products, meat and seafood products, and herbal cooking oils. The symptoms are dry mouth, double vision, muscular paralysis, respiratory failure, nausea, vomiting, and diarrhea. Botulism can be fatal and requires immediate medical attention. The incubation period is usually from 4 to 36 hours after eating contaminated food. The recovery period can extend from a week to a full year. Cooking food properly is imperative.
- *Clostridium perfringens*: This produces heat-stable spores that prevail in undercooked foods and foods left unrefrigerated and at room temperature. The sources are meat and meat products. The symptoms are abdominal pain and diarrhea with occasional vomiting and nausea.
- *Cryptosporidium parvum*: This protozoan causes illness from contaminated water, fecal-to-oral contamination, and raw and undercooked foods. Sources of contamination are water, milk, contaminated food, and person-to-person transmission. The incubation period is usually 2 to 10 days. Symptoms consist of diarrhea, gastrointestinal cramps, and loss of appetite. Prevention can be achieved with good hygiene and boiling potentially contaminated water.
- Pathogenic *Escherichia coli*, O157:H7: Deadly toxins can be produced. The symptoms are hemorrhagic colitis. Sources are raw or undercooked meat, raw milk, unpasteurized juice, and contaminated water. The symptoms are severe cramps, hemorrhagic colitis, and bloody or nonbloody diarrhea. The incubation period is usually 3 to 4 days and may last from 5 to 8 days.
- *Listeria monocytogenes*: Illness or death in pregnant women, fetuses, and newborns may occur. The sources are meat, seafood, poultry, and unpasteurized dairy or dairy products. The symptoms are fever, muscle aches, headaches, nausea, vomiting, diarrhea, meningitis, and miscarriages. The incubation period can be from as little as 2 days to 3 weeks. Sanitary and hygiene practices are important.
- *Giardia lamblia*: This is a protozoan parasite. The disease is termed giardiasis. The symptoms are watery stool, diarrhea, stomach cramps, and lactose intolerance. The protozoan is found in soil water and foods that have been contaminated with the feces of animals or infected humans. Good hygiene and sanitization are preventive measures.
- *Norovirus* (Norwalk, Norwalk-like): This virus can cause nausea, vomiting, diarrhea, fever, and abdominal cramps. The sources are shellfish, oysters, salads, frosting, and person-to-person contamination. The incubation

period is usually 24 to 48 hours but can occur as early as 12 hours and can last for 12 to 60 hours.
- *Salmonella* serotype Enteritidis: This *Salmonella* causes salmonellosis. Contamination can be from eggs, poultry, meat, or milk products. The symptoms can be diarrhea, fever, vomiting, nausea, and abdominal cramps. The incubation period is usually 12 to 72 hours after ingesting contaminated food and can last for 4 to 7 days.
- *Shigella*: *Shigella* is a bacterium that results from poor hygiene, water contaminated by human feces, and unsanitary food handling. It can cause dysentery. The sources can be salads, raw vegetables, dairy products, raw oysters, ground beef, and poultry. The incubation period is 12 to 60 hours, but often is as long as 7 days. The duration of illness is usually 5 to 7 days.
- *Staphylococcus aureus*: The bacterium is found on the skin and in nasal passages. As a result of poor hygiene and sanitation, it can be transferred from person to person. Sources can be dairy products, raw meats, poultry, and humans. Following surgical operations, it has caused infection. The symptoms are abdominal cramps, nausea, vomiting, and diarrhea. Symptoms can appear within 30 minutes to 8 hours after ingesting contaminated food. The duration of infection is usually 24 to 48 hours.
- *Vibrio cholerae*: The bacterium is found in estuary environments. It can result in death. The primary sources are raw and undercooked seafood as well as contaminated water and food. Potential symptoms are diarrhea, vomiting, and leg cramps. The incubation period can occur from 5 to 6 days following intake of contaminated food. The duration is usually 7 days.
- *Vibrio parahaemolyticus*: This bacterium resides in saltwater and can cause gastroenteritis. The sources can be raw or undercooked fish or shellfish. The symptoms can be fever, nausea, vomiting, headache, chills, and stomach cramps. The incubation period can last from 4 to 96 hours and can last for 2.5 days following consumption of contaminated food.
- *Vibrio vulnificus*: This also inhabits saltwater. It can enter through an exposed wound. The sources are raw fish and shellfish, especially oysters. The symptoms can manifest in diarrhea, nausea, gastrointestinal pain, fever, vomiting, chills, and in some cases sores or blisters on a person's legs. The incubation period is usually 16 hours after ingesting contaminated food or exposure to contaminated water. The infection may last from 2 to 3 days.
- *Yersinia enterocolitica*: It is the cause of the disease yersiniosis. It can mimic acute appendicitis. The primary sources are raw meat and seafood, dairy and dairy products, as well as contaminated water. The predominant symptoms are vomiting, fever, diarrhea, and stomach pain. The symptoms can be severe in children. Symptoms may appear in 1 to 2 days after ingestion and last from 1 to 2 days.

Table 10.1 provides data on reported and estimated illnesses, hospitalization rates, and case fatality rates for known foodborne bacterial pathogens in the United

TABLE 10.1
Reported and Estimated Illnesses, Hospitalization Rates, and Case Fatality Rates for Known Foodborne Bacterial Pathogens in the United States

Pathogen	Estimated Total Cases	Hospitalization Rate	Case Fatality Rate
Bacterial			
Bacillus cereus	27,360	0.006	0.0000
Botulism	58	0.800	0.0769
Brucella spp.	1,554	0.550	0.0500
Campylobacter spp.	2,453,926	0.102	0.0010
Salmonella, nontypical	1,412,498	0.221	0.0078
Escherichia coli			
O157:H7	73,480	0.295	0.0083
Non-O157:H7 STEC	36,740	0.295	0.0083
Enterotoxigenic	79,420	0.005	0.0001
Other diarrheagenic	79,420	0.005	0.0001
Listeria monocytogenes	2,518	0.922	0.2000
Salmonella typhi	824	0.750	0.0040
Shigella spp.	448,240	0.139	0.0016
Staphylococcus			
Food poisoning	185,060	0.180	0.0002
Foodborne	50,920	0.130	0.0000
Vibrio cholerae	54	0.340	0.0060
Vibrio vulnificus	94	0.910	0.3900
Vibrio, other	7,880	0.126	0.0250
Yersinia enterocolitica	96,368	0.242	0.0005
Clostridium perfringens	248,520	0.242	0.0005
Brucella species	1,554	0.550	0.0500
Subtotal	5,204,934		
Parasitic			
Cryptosporidium parvum	300,000	0.150	0.0050
Cyclospora cayetanensis	16,264	0.020	0.0005
Giardia lamblia	2,000,000	Not available	Not available
Toxoplasma gondii	225,000	Not available	Not available
Trichinella spiralis	52	0.081	0.0030
Subtotal	2,541,316		
Viral			
Norwalk-like virus	23,000,000	Not available	Not available
Rotavirus	3,900,000	Not available	Not available

(Continued)

TABLE 10.1 (Continued)
Reported and Estimated Illnesses, Hospitalization Rates, and Case Fatality Rates for Known Foodborne Bacterial Pathogens in the United States

Pathogen	Estimated Total Cases	Hospitalization Rate	Case Fatality Rate
Astrovirus	3,900,000	Not available	Not available
Hepatitis A	83,391	0.130	0.0030
Subtotal	30,883,391		
Total	38,629,641		

Source: Data from Centers for Disease Control and Prevention (CDC). 1999. *CDC's Emerging Infections Programs—1999 Surveillance Results.* Atlanta, GA: Centers for Disease Control and Prevention; Mead PS, Slutsker L, Dietz V, McCaig LF, Bresee JS, Shapiro G, Griffin PM, and Tauxe RV. 1999. Food-related illnesses and death in the United States. *Emerg Infect Dis* 5(5): 607–625.

States. It was estimated that foodborne diseases caused approximately 76 million illnesses, 325,000 hospitalizations, and 5,000 deaths in the United States each year (Mead et al. 1999).

The Centers for Disease Control and Prevention (CDC) indicate that some portion of gastrointestinal illnesses is caused by foodborne agents not yet identified (Mead et al. 1999). Deaths are also unreported. And, Mead et al. (1999) indicated that food-related death information is especially difficult to obtain because pathogen-specific surveillance systems rarely collect information on illness outcome.

Table 10.2 provides data on cases and incidence rates of foodborne diseases each year in the United States (Swartz 2002.). Obviously, there are many more cases throughout the United States. Furthermore, many cases are not reported by doctors.

Note that the source of data was the CDC (1999). Surveillance occurred in eight states (Connecticut, Georgia, Minnesota, and Oregon and selected counties in the states of California, Maryland, New York, and Tennessee) through more than 300 clinical laboratories. The total population assessed was 25.6 million.

Table 10.3 provides data on the percentage of people hospitalized in the United States as a result of foodborne pathogens (CDC 1999). The data were obtained from the National Ambulatory Medical Care Survey (McCaig and Burt 1999).

The CDC (2011) indicated that foodborne illness can be prevented.

In subsequent data, the CDC indicated that there were two major groups of foodborne illnesses. There were 31 known pathogens and several unspecified agents. The 31 known foodborne pathogens caused an estimated annual 9.4 million illnesses (6.6–12.7 million), resulting in 1,351 (712–2,268) deaths. The unspecified number of agents resulted in 38.4 (19.8–61.2) million estimated annual illnesses (Centers for Disease Control and Prevention (CDC) 2011).

TABLE 10.2
Cases and Incidence Rates of Foodborne Diseases Each Year in the United States

Pathogen	Cases	Incidence Rate per 100,000
Bacteria		
Salmonella	4,533	17.7
Campylobacter	3,794	14.8
Shigella	1,031	4.0
Escherichia coli O157:H7	530	2.0
Yersinia	163	0.6
Listeria	113	0.5
Vibrio	45	0.2
Total	10,209	
Parasites		
Cryptosporidium	474	1.5
Cyclospora	14	0.04
Total	488	

Source: Swartz MM. 2002. Human diseases caused by foodborne pathogens of animal origin. *Clin Infect Dis* 34(Suppl 3): S111–S122.

TABLE 10.3
Percentage of People Hospitalized Each Year in the United States as a Result of Infections from Foodborne Pathogens

Percentage	Percentage Hospitalized
Listeria	88
Escherichia coli O157:H7	37
Yersinia	36
Vibrio	25
Salmonella	22
Shigella	14
Campylobacter	11

Source: Centers for Disease Control and Prevention (CDC). 1999. *CDC's Emerging Infections Programs—1999 Surveillance Results.* Atlanta, GA: Centers for Disease Control and Prevention.

The top five pathogens causing the most illness annually were

- *Norovirus*: 5,461,731 illnesses (range 3,227,078–8,309,480)
- *Salmonella* (nontyphoidal): 1,027,561 (range 644,786–1,679,667)
- *Clostridium perfringens*: 965,958 (range 192,316–2,483,309)
- *Campylobacter* spp.: 845,024 (range 337,031–1,611,083)
- *Staphylococcus aureus*: 241,148 (range 72,341–529,417)

The top five pathogens contributing acquired foodborne illnesses resulting in estimated deaths were

- *Salmonella* (nontyphoidal): 378 estimated deaths
- *Toxoplasma gondii*: 327 estimated deaths
- *Listeria monocytogenes*: 255
- *Norovirus*: 149
- *Campylobacter*: 76

These numbers above estimated 88% of the deaths as a result of foodborne illnesses.

Thomas et al. (2013) reported on the estimates of foodborne illnesses in Canada for 30 specified pathogens and unspecified agents based on data from 2000 to 2010 using a population census of 2006. It was estimated that each year there were 1.6 million (1.2–2.0 million) illnesses related to 30 known pathogens and 2.4 million (1.8–3.0 million) related to unspecified agents. The leading pathogens accounting for 90% of the illnesses were *Norovirus*, *Clostridium perfringens*, *Campylobacter* spp., and nontyphoidal *Salmonella* spp. These accounted for 90% of the pathogen-specific foodborne illnesses. At that time, approximately 1 in 8 Canadians experienced an episode of domestically acquired foodborne illness each year in Canada (Thomas et al. 2013).

In developing countries, the situation is more acute. In addition to the cited organisms in the United States and Canada, *Vibrio cholerae* is a major public health concern for the cholera. This results from both water and food contamination.

Other foodborne illnesses and infection have been caused by enterohemorrhagic *Escherichia coli* O157:H7 and *Listeria*, which causes listeriosis. The illness among infants and the elderly may be severe and even fatal.

It was indicated that there were other sources of foodborne pathogens as unspecified agents. These could refer to naturally occurring toxins such as mycotoxins, marine biotoxins, cyanogenic glycosides, and toxins produced by mushrooms. Mycotoxins, such as aflatoxin and ochratoxin A, are found in numerous staple foods.

Prevention and control of foodborne diseases need to be achieved and reduced by farmers, producers, and the public. Organic farmers using manure need to use properly composted materials. Other farmers need to prevent domestic and wild animals from entering fields and contaminating crops. Where possible, fruits and vegetables sold to the public should be washed thoroughly. Good hygienic practice must be used by the public. Water supplies need to be protected from animal and human waste contamination.

WATERBORNE DISEASES

As stated previously, many of the foodborne diseases are also related to waterborne organisms. There are other direct and indirect diseases associated with water courses; for example, malaria is transmitted via a bite by the female *Anopheles* mosquito. With the bite, the parasitic protozoan is introduced into the human circulatory system. The mosquito breeds in shallow stagnant water. Most of the waterborne diseases caused by pathogenic microorganisms are transmitted from freshwater sources. These can be bathing, swimming, drinking, and washing water. Food can be contaminated by the water.

The major waterborne diseases occur in developing countries as the result of lack of sanitary conditions, lack of infrastructure to remove and treat wastes, lack of or poor hygiene, poor education, and stagnant water sources.

Table 10.4 presents waterborne pathogens confirmed by epidemiological studies and case histories. These studies were done with healthy adult volunteers and as such apply to only a portion of the exposed population.

Bacteria and protozoa are the predominant microorganisms causing waterborne diseases. These organisms enter the intestinal tract and then invade tissues or the circulatory system. Viruses are also causes of waterborne diseases. Metazoan parasites (e.g., roundworms or nematodes such as *Dracunculus*, which causes guinea worm disease) come from swallowing water in which there are certain copepods, which are small crustaceans found in the sea and nearly every freshwater habitat and act as vectors for nematode *Dracunculus*.

Some members of the Schistosomatidae family infect individuals who have skin contact with infected water. The disease schistosomiasis affects hundreds of millions of people worldwide.

Table 10.5 shows some waterborne infections, the agent, and the source the agent in the water supply, and general disease symptoms. Many of the symptoms are similar (e.g., diarrhea), and the sources can be identified by specific tests. Particularly with respect to developing countries, where proper sanitation is not available, fecal contamination (e.g., fecal to oral) is a major source of infection. Runoff containing fecal matter will contaminate drinking and bathing water sources.

Although the United States has one of the safest drinking water systems in the world, there are an estimated 4 to 32 million cases of acute gastrointestinal illness (AGI) per year from public drinking water systems. This estimate does not include waterborne illness from nonpublic drinking water systems (e.g., private wells), recreational water, or water for other uses (e.g., irrigation, medical uses, or building water systems). The frequency of disease from all water exposures is likely higher, but the overall prevalence of waterborne illness in the United States is unknown (CDC 2013).

TABLE 10.4
Waterborne Pathogens and Their Significance in Water Supplies

Pathogen	Health Significance[a]	Persistence in Water Supplies[b]	Relative Infectivity[c]
Bacteria			
Burkholderia pseudomallei	High	May multiply	Low
Campylobacter jejuni, Campylobacter coli	High	Moderate	Low
Escherichia coli: Pathogenic	High	Moderate	Low
Escherichia coli: Enterohemorrhagic	High	Moderate	Low
Legionella spp.	High	May multiply	Low
Nontuberculous mycobacteria	Low	May multiply	High
Pseudomonas aeruginosa	Moderate	May multiply	Moderate
Salmonella typhi	High	Moderate	Low
Other salmonellae	High	May multiply	Low
Shigella spp.	High	Short Low	High
Vibrio cholerae	High	Short to long	Low
Yersinia enterocolitica	High	Long	Low
Viruses			
Adenoviruses	High	Long	High
Enteroviruses	High	Long	High
Astroviruses	High	Long	High
Hepatitis A virus	High	Long	High
Hepatitis E virus	High	Long	High
Noroviruses	High	Long	High
Sapoviruses	High	Long	High
Rotavirus	High	Long	High
Protozoa			
Acanthamoeba spp.	High	May multiply	High
Cryptosporidium parvum	High	Long	High
Cyclospora cayetanensis	High	Long	High
Entamoeba histolytica	High	Moderate	High
Giardia intestinalis	High	Moderate	High
Naegleria fowleri	High	May multiply	Mod
Toxoplasma gondii	High	Long	High
Helminth			
Dracunculus medinensis	High	Moderate	High
Schistosoma spp.	High	Moderate	High

Source: World Health Organization (WHO). 2007. http://www.who.int/water_sanitation_health/gdwqrevision/watborpath/en

[a] Health significance relates to the severity of impact, including associations with outbreaks.
[b] Detection period for infective stage in water at 20°C; short, up to 1 week; moderate, 1 week to 1 month; long, greater than 1 month.
[c] From experiment with human volunteers, from epidemiological evidence and animal studies. High mean infective dose can be $1–10^2$ organisms or particles, moderate can be $10^2–10^4$, and low is greater than 10^4.

TABLE 10.5
Some Waterborne Infections, the Agents Involved, Source of Water Supply, and General Disease Symptoms

Disease and Transmission	Microbial Agent	Sources of Agent in Water Supply	General Symptoms
Bacteria			
Botulism Contaminated food	*Clostridium botulinum* Vomiting, diarrhea, difficulty swallowing, respiratory failure resulting in death	Contaminated water	
Campylobacteriosis	Most commonly caused by *Campylobacter jejuni*	Fecal-contaminated drinking water	High fever, dysentery symptoms
Cholera	*Vibrio cholerae*	Contaminated drinking water	Watery diarrhea, nausea, vomiting, rapid pulse, hypovolemic shock resulting in death
Escherichia coli infection	Certain strains	Contaminated water	Diarrhea, prolonged dehydration
Mycobacterium	*Mycobacterium marinum*	Exposure in swimming pools	Lesions on elbows, knees, feet, and hands
Dysentery	Species of *Salmonella* and *Shigella*, *Shigella dysenteriae*	Contaminated water	Feces with blood or mucus, bloody vomit
Legionellosis	*Legionella pneumophila*	Contaminated water	Influenza, pneumonia, fever, chills, ataxia, anorexia, muscle aches
Leptospirosis	*Leptospira*	Animal urine-contaminated water	Flu-like symptoms, meningitis, liver damage, and renal failure
Salmonellosis	*Salmonella*	Food, contaminated water	Diarrhea, fever, vomiting, abdominal cramps
Typhoid fever	*Salmonella typhi*	Feces-contaminated water	High fever, delirium, enlargement of spleen and liver
Protozoa			
Amoebiasis	*Entamoeba histolytica*	Sewage, contaminated drinking water	Fever, abdominal pain, diarrhea, bloating
Cryptosporidiosis	*Cryptosporidium parvum*	Animal manure	Flu-like symptoms, diarrhea, weight loss, nausea
Giardiasis	*Giardia lamblia*	Water contamination	Diarrhea, abdominal pain, bloating, flatulence

(Continued)

TABLE 10.5 (Continued)
Some Waterborne Infections, the Agents Involved, Source of Water Supply, and General Disease Symptoms

Disease and Transmission	Microbial Agent	Sources of Agent in Water Supply	General Symptoms
Parasites			
Schistosomiasis	*Schistosoma*	Water contaminated with certain snails	Blood in urine, fever, chills, rash
Dracunculiasis	*Dracunculus medinensis*	Stagnant water containing larvae	Rash, vomiting, diarrhea, asthmatic attack
Taeniasis	*Taenia* (tapeworms)	Contaminated drinking water	Intestinal discomfort, weight loss
Ascariasis	*Ascaris lumbricoides*	Feces-contaminated drinking water	Inflammation, fever, diarrhea
Viruses			
Severe acute respiratory syndrome (SARS)	Coronavirus	Poorly treated water	Fever, myalgia, gastrointestinal discomfort, cough, and sore throat
Hepatitis A	Hepatitis A virus	Water and food	Fever, fatigue, abdominal pain, diarrhea, weight loss, jaundice, depression

Source: Waterborne diseases. n.d. http://en.wikipedia.org/wiki/waterborne_diseases.

WHO states that water-related diseases include:

- Those due to microorganisms and chemicals in water people drink
- Diseases like schistosomiasis that have part of their life cycle in water
- Diseases like malaria with water-related vectors
- Drowning and some injuries
- Others, such as legionellosis, carried by aerosols containing certain microorganisms

Obviously, not all of these result in gastrointestinal diseases. However, many of the illnesses of children in developing countries that result in diarrhea and in many cases death are due to gastrointestinal illness from waterborne causes. Many of these could be prevented through good hygiene and eliminating sources (e.g., malaria).

Diarrhea occurs worldwide and causes 4% of all deaths and 5% of health loss to disability. It is most commonly caused by gastrointestinal infections, which kill around 2.2 million people globally each year, mostly children in developing countries. The use of water in hygiene is an important preventive measure, but contaminated water is also an important cause of diarrhea. Cholera and dysentery cause severe, sometimes life-threatening, forms of diarrhea.

SUMMARY

Foodborne and waterborne pathogens can contaminate individuals through food, water, soil, and air. These pathogens result in large numbers of illnesses and deaths. The burden to society is in the billions of dollars.

In developed countries, as a result of better hygiene, sanitation, and education, the illnesses, hospitalizations, and death are considerably less than in developing countries. However, the numbers and costs are still high. In developing countries, mostly children are killed.

Water contamination can occur as a result of human and animal fecal matter. In developing countries, especially in villages, drinking water resources are below the deposition of human wastes.

The agricultural industry in the United States particularly needs to prevent contamination by workers as well as potential contamination from runoff containing animal waste. Many agricultural workers migrating from warm or hot climates can infect crops, especially during harvesting. Education to prevent contamination by workers is essential. Workers should also have adequate and clean facilities providing good hygienic practices.

The global burden, human toll, and costs could be greatly reduced by better education, improved sanitation (especially in developing countries), and hygiene.

REFERENCES

Batz MB, Hoffman S, and Morris JG, Jr. Ranking the Risks: The 10 Pathogen-Food Combinations with the Greatest Burden on Public Health. Gainesville, FL: University of Florida, Emerging Pathogens Institute.

Centers for Disease Control and Prevention (CDC). 1999. *CDC's Emerging Infections Programs—1999 Surveillance Results*. Atlanta, GA: Centers for Disease Control and Prevention.

Centers for Disease Control and Prevention (CDC). 2011. Vital signs: incidence and trends of infection with pathogens transmitted commonly through food. *MMWR Morb Mortal Wkly Rep* 60: 749–755.

Centers for Disease Control and Prevention (CDC). 2011. CDC Estimates of Foodborne Illness in the United States. http://www.cdc.gov/foodborneburden/2011-foodborne-estimates.html

Centers for Disease Control and Prevention (CDC). 2013. Magnitude and burden of waterborne disease in the U.S. http://www.cdc.gov/healthywater/burden/index.html.

Epstein, E. 2011. *Industrial Composting*. Boca Raton, FL: CRC Press, Taylor and Francis.

Foodborne illness. n.d. http://en.wikipedia.org/wiki/Foodborne_illness

McCaig LF and Burt CW. 1999. National Hospital Ambulatory Medical Care Survey: 1999 Emergency Department Summary. *Adv Data* Jun 25 (320): 1–34.

Mead PS, Slutsker L, Dietz V, McCaig LF, Bresee JS, Shapiro G, Griffin PM, and Tauxe RV. 1999. Food-related illnesses and death in the United States. *Emerg Infect Dis* 5(5): 607–625.

National Institute of Environmental Health Sciences (NIEHS). 2010. A Report Outlining the Research Needs on the Human Effects of Climate Change. http://www.niehs.nih.gov/health/materials/a_human_health_perspective_on_climate_change_full_report_508.pdf

Swartz MM. 2002. Human diseases caused by foodborne pathogens of animal origin. *Clin Infect Dis* 34(Suppl 3): S111–S122.

Thomas MK, Murray R, Flockhart L, Pintar F, Pollari F, Fazil A, Nesbitt A, and Marshall B. 2013. Estimates of the burden of foodborne illnesses in Canada for 30 specified pathogens and unspecified agents, circa 2006. *Foodborne Pathog Dis* 10: 639–648.

US Food and Drug Administration. 2014. Food safety for moms-to-be: medical professionals—foodborne pathogens. http://www.fda.gov/food/resourcesforyou/healtheducators/ucm091681.htm

Waterborne diseases. n.d. http://en.wikipedia.org/wiki/Waterborne_disease

World Health Organization (WHO). 2003. *Guidelines for Drinking Water Quality*, 3rd edition. Geneva, Switzerland: WHO.

World Health Organization (WHO). 2003. Hazard characterization for pathogens in food and water. Microbial health risk assessment.

World Health Organization (WHO). 2007. Rolling revision of the WHO guidelines for drinking-water quality. *Waterborne Pathogens*. http://www.who.int/water_sanitation_health/gdwqrevision/watborpath/en

11 Disposal and Management of Solid Waste

INTRODUCTION

The disposal and management of solid waste will generally differ greatly in developed counties than in developing countries. This is partly the result of the absence of infrastructure in developing countries. In developed countries, the presence of wastewater-handling facilities; septic systems for single homes (i.e., disposal in the absence of sewerage); trash or garbage pickup; incineration with heat recovery; and recycling and recovery of paper, metals, plastic, and glass are part of a regulated and controlled infrastructure.

In many developing countries, disposal of both human wastes and solid waste is unhygienic, resulting in proliferation of pathogens and diseases. There are opportunities to reduce these conditions through low-cost biological systems and in some cases chemical systems.

The most opportune biological system is composting (i.e., controlled decomposition of organic wastes). In many cases, the compost can be used as an organic fertilizer or for plant nutrients to produce food. If lime is available, especially where soils are acidic as in the case of humid areas, it is both an excellent disinfectant and a source of calcium to plants.

The disposal options are

- Landfill or dumps
- Incineration or burning
- Anaerobic digestion
- Land application
- Composting
- Lime stabilization

I have separated land application and composting as two distinct entities even though many agricultural and horticultural scientists would consider composting as a subunit of land application. However, compost can also be used as a mulch to minimize runoff and erosion, thus not directly incorporated into the soil. Furthermore, dried compost can be used as a source of heat.

LANDFILLS OR DUMPS

Landfills today, especially in developing countries, are designed to reduce environmental impacts, such as groundwater contamination and air pollution. Groundwater contamination can principally result from both inorganic and organic chemical contaminants. Inorganic contaminants could be nitrogen and phosphorus from fertilizers, whereas organic contaminants could be pesticides, surfactants, lubricants, and similar compounds.

The principal air pollution from landfills is from the discharge of the methane that results from organic material decomposition under anaerobic conditions. Methane is 25 times the air pollutant compared to carbon dioxide. Methane is also an explosive gas and has been known to seep into basements in nearby residences. There have also been incidents of explosions when buildings were built on former landfills. An example is one that occurred in the Boston, Massachusetts, area; newly developed low-cost housing had to be abandoned as a result of methane seepage into basements and an explosion.

Dumps are places for indiscriminant disposal of human and other solid wastes. They are a source of pathogens both through contamination of humans who scavenge the dump and via air pollution. As pointed out previously in this book, not only are the scavengers subject to diseases but also they bring diseases into their households and communities. Today, in addition to traditional garbage, a large amount of electronic hardware is disposed. This can result in heavy metal and some organic contamination.

Singh, Datta, and Nema (2007) reported that in India uncontrolled land disposal of municipal solid waste (MSW) is common. They noted that groundwater contamination is the most important source of contamination. Financial constraints are the predominant restraint for improvements (Singh, Datta, and Nema 2007). In India in 2007, it was reported that nearly 90% of the MSW was disposed in open dumps and landfills, creating public health and environmental issues (Sharholy et al. 2008). Uncontrolled dumps and landfills are also common in many other developing countries in Asia, Africa, and South America. As indicated previously, open dumps and landfills are scavenged for usable items. However, the only possible management of these dumps and landfills is soil covering on a daily basis. Groundwater contamination can be avoided by the use of liners. These actions then become a financial issue.

INCINERATION AND BURNING

Incineration is controlled burning in an enclosed facility. Incineration can be accomplished with energy recovery. Air pollution and disposal of ash can be regulated to minimize environmental impacts. Burning of dumps can be deliberate or the result of methane emission. Workers can initiate a fire by the careless disposal of cigarettes. The United States Environmental Protection Agency (USEPA) reports that the first incinerator was built in New York in 1885. By the middle of the twentieth century, there were hundreds of combustion facilities. Combustion of MSW grew in the 1980s, with more than 15% of all US MSW combusted by the early 1990s. The majority of the nonhazardous waste incinerators was recovering energy by this time and had installed pollution control equipment. Following incineration, there is still ash to be disposed. The ash is often toxic, containing heavy metals such as vanadium, manganese, chromium, nickel, cadmium, arsenic, mercury, and lead. There has been an attempt to use the ash for road building where it can be sealed to prevent leaching. Alternative technologies are available or in development, such as composting or anaerobic digestion with energy recovery, autoclaving, or mechanical heat treatment using steam.

Incinerators are expensive and require control of emissions. However, heat recovery is possible and may result in financial feasibility. Most likely, developing countries would not opt for this way to manage MSW. In developed countries, especially Europe, this may be a feasible option, providing the incinerators can significantly reduce air pollution.

ANAEROBIC DIGESTION

Anaerobic digestion is the biological decomposition of organic waste by bacteria without oxygen. This can result in the production of combustible methane gas that can be used for heating or operation of engines. Anaerobic digestion is common at wastewater treatment plants in the United States, Canada, and Europe. It is not common in the use of MSW. Anaerobic digestion of solid waste is more complicated. The material is heterogeneous, often containing metal and plastics, which are not digested. Also, the moisture content is low; however, this can be adjusted.

Generally, anaerobic digesters in Europe and America are large and expensive structures. However, I have seen modest, low-cost units built from plywood and used to decompose food waste in California and Hawaii. These could be designed in small units for villages in developing countries and used to provide hot water and even cooking for better hygiene. The residue can be used as a fertilizer and land applied (see land application).

In India, inside a poor home I saw a simple animal waste anaerobic digester used for cooking. My point is that there are opportunities in developing countries that could provide for better methods of disposal of waste.

The efficiency of anaerobic digestion of wastes to kill human and animal pathogens is limited. Since anaerobic digested wastes, especially human wastes, are land applied, the efficiency of anaerobic digestion on pathogen destruction is discussed in the next section on the land application of wastes.

LAND APPLICATION

Direct land application of human and animal wastes has been used for many centuries. Johansson et al. (2005) discussed the potential risks when spreading anaerobic digested residues on grass silage crops. One of the most comprehensive reviews regarding the sources and fate of pathogens during land application of wastes was reported by Gerba and Smith (2005). They found that more than 150 known enteric pathogens might be present in untreated wastes. Use of untreated or raw wastes for fertilization of food crops is not recommended. They can be used as fertilizer for trees or crops not to be eaten by humans. There are regulations on land application of wastes in the United States, Canada, and Europe. In the United States, the USEPA regulation regarding the various processes that can be used to reduce pathogens prior to land application are cited in 40 CFR 503 (processes to significantly reduce pathogens [PSRPs]) (USEPA 1995).

Prior to land application, the waste can be disinfected using anaerobic digestion, composting, or lime stabilization. Human wastes should not be applied near water wells used for drinking or bathing. They should be applied downstream from human habitation. Human wastes applied to land should be incorporated into the soil as soon as possible. Children should not be allowed to play in the area. Workers should change clothes prior to returning home or contacting other persons.

COMPOSTING

Composting of human wastes is being done in United States, Canada, and Europe. Facilities range from very small to large. The two largest facilities are in California and Canada. In developing countries, composting can be done on a very small scale as well as for large municipalities. The technology selected depends on location, material, land availability, power availability, human resources, and financial resources. Location refers to proximity to homes and soil surface conditions. A compacted surface may be needed if heavy equipment is used. Regardless of technology, some type of bulking agent is required. This could be straw, grass, bagasse, shredded wood or tree branches, nut hulls, or any other type of local material. The amount of land required depends on the technology selected. Windrow technology would require more land than the aerated static pile. Power is needed in the form of fuel for equipment or electricity for blowers for the aerated pile system. Both solar and wind power could be used for motors on blowers for the aerated pile system. No power is needed for the system devised by Sir Albert Howard, an English botanist known for his refinement of a traditional Indian composting system (Van Vuren 1949). A simple method for composting human or animal wastes, also based on studies by Howard, is shown in Figure 11.1 (Howard 1935, 1936, 1943; Epstein 1997).

For example, the aerated pile system basically only requires a blower. The piles can be constructed by humans using shovels or a front-end loader. Pipes to provide the aeration range from bamboo to drainage plastic pipes or iron pipes. The aerated pile system has been used in climates where rainfall exceeded 100 cm per month but can be used where rainfall is much higher. The use of fabric covers allows the piles to be constructed in virtually any climate.

Windrow composting can also be used with minimal equipment. Turning can be accomplished by humans using shovels.

The disposal and management of various wastes is feasible by composting for either small or large communities. In developing countries, the collection of various wastes harmful to humans can be composted by simple processes and the residue used as a fertilizer. If during the composting process the temperature were allowed to reach 55°C for several days, then disinfection would occur. Pathogen destruction is a function of time and temperature. The higher the temperature is, the less time is needed for disinfection. Thus, in developing countries the wastes can be held in piles for long periods of time.

Disposal and Management of Solid Waste

FIGURE 11.1 A simple process for composting human or animal wastes.

LIME STABILIZATION

The addition of lime can be an effective means for pathogen destruction. Most pathogens strive at neutral to slightly acid conditions. Bean et al. (2007) state that liming is a cost-effective treatment currently employed in much USEPA class B biosolids production.

Lime stabilization is a simple process. Its advantages are simplicity and low cost. The USEPA stated that the waste must reach a pH of 12 for 2 hours to be effective in reducing pathogens. During lime treatment, ammonia, which is a disinfectant, is released. Furthermore, heat is also produced. This combination of ammonia and heat release can result in pathogen destruction.

For example, India has an abundance of lime primarily used for cement production. Some of this lime could effectively be used for waste disinfection.

SUMMARY

Waste disposal is a major problem in developing countries. Solid waste production, accumulation, and disposal are major sources of diseases.

There are several potential disposal methods:

- Landfill or dumps
- Incineration or burning
- Anaerobic digestion
- Land application
- Composting
- Lime stabilization

Landfills that are properly constructed and operated, land application, composting, and lime stabilization are the least-expensive options. Landfills can be designed to capture methane for heating and other uses. It is an affective disposal method for all types of human wastes. Its limitation is that if energy is not recovered and used, the technology results in solid waste disposal without any benefits. Furthermore, the design must consider groundwater contamination.

Incineration with heat recovery is expensive, and air pollution is a major consideration. The use of air pollution devices makes the technology even more expensive. The technology has a primary use in developed countries that can afford it.

Certain wastes (e.g., human wastes), when land applied, are fertilizers for crops that are not to be eaten raw. Land application is not feasible for MSW (e.g., plastics, paper, and other materials).

Composting when temperatures reach 55°C can be effective in pathogen destruction, and the product can be used as a fertilizer and soil conditioner. However, the technology today is primarily used for human and animal wastes. This also applies for lime stabilization. These technologies are not suitable for MSW consisting of plastics, metals, and other materials.

It is evident that several low-cost and implementable technologies are available and suitable for developing countries.

REFERENCES

Bean CL et al. 2007. Class B alkaline stabilization to achieve pathogen inactivation. *Int J Environ Res Public Health*. 4(1): 53–60.
Epstein E. 1997. *The Science of Composting*. Lancaster, PA: Technomic.
Gerba CP, and Smith JE Jr. 2005. Sources of pathogenic microorganisms and their fate during land application of wastes. *J Environ Quality* 34: 42–48.
Howard A. 1935. The manufacture of humus by the Indore process. *J R Soc Arts* 74: 26–60.
Howard A. 1936. Manufacture of humus by the Indore process. *Nature* 137(3461): 363.
Howard A. 1943. *An Agricultural Testament*. New York: Oxford University Press.
Johansson M, Emmoth E, Salomonsson AC, and Albihm A. 2005. Potential risks when spreading anaerobic digestion residues on grass silage crops—survival of bacteria, moulds and viruses. *Grass Forage Sci* 60: 175–185.
Sharholy M, Ahmad K, Mahmood G, and Trivedi RC. 2008. Municipal solid waste management in Indian cities. *Waste Manage* 28: 459–467.
Singh RK, Datta M, and Nema AK. 2007. Ground water contamination hazard potential rating of municipal solid waste dumps and landfills. In *Proceedings of the International Conference on Sustainable Solid Waste Management, Chennai, India*, 296–303.
United States Environmental Protection Agency (USEPA). 1995. *A Guide to the Biosolids Risk Assessments for the EPA Part 503 Rule*. Washington, DC: US Environmental Protection Agency.
Van Vuren JPJ. 1949. *Soil Fertility and Sewage*. London: Faber and Faber.

Appendix: Details of the Pathogens and Their Diseases

ACTINOBACILLUS PLEUROPNEUMONIAE

Actinobacillus pleuropneumoniae causes porcine pleuropneumonia, a highly contagious disease for which there is no effective vaccine. It is found in most major swine-producing regions. It is a pathogen in mammals, bird, and reptiles.

SOURCES

en.wikipedia.org/wiki/Actinobacillus_pleuropneumoniae.
http://www.ncbi.nlm.nih.gov/pubmed/11880056.

ACINETOBACTER BAUMANNII

Acinetobacter baumannii is considered one of the most difficult Gram-negative bacilli to control and treat. It can cause outbreaks of infection, including meningitis, urinary infections, pneumonia, and wound infections. It is found in soil and water. It is an opportunistic nosocomial pathogen. Therefore, it poses little risk to healthy persons. It can live on the skin and survives for extensive periods under a variety of environmental conditions.

SOURCES

Antunes LCS, Visca P, and Towner KJ. 2014. *Acinetobacter baumannii*: evolution of a global pathogen. *Pathog Dis* doi: 10.1111/2049-632X.12125.
Maragakis LL and Perl TM. 2008. *Acinetobacter baumannii*: epidemiology, antimicrobial resistance, and treatment options. *Clin Infect Dis* 46(8): 1254–1263.

ACTINOMYCES SPP.

Actinomycetes are a specific group of bacteria. Morphologically, they resemble fungi because of their elongated cells that branch into filaments or hyphae.

During composting, reaching high temperatures, thermophilic and thermotolerant actinomycetes are responsible for decomposition of the organic matter. In the initial phase of composting, the intensive increase of microbial activity leads to self-heating of the organic material. High temperatures in composting help to kill viruses, pathogenic bacteria (e.g., coliforms), parasites, and weed seeds. Actinomycetes live predominantly aerobically; that is, they need oxygen for their metabolism. Aeration during composting is therefore important. During the composting process, the actinomycetes degrade natural substances such as chitin or cellulose.

Some thermophilic and thermotolerant actinomycetes are found to be responsible for allergic symptoms in the respiratory tract (e.g., "extrinsic allergic alveolitis").

SOURCES

en.wikipedia.org/wiki/Actinobacillus_pleuropneumoniae

Kokotovic B, Angen O, and Bisgaard M. 2011. Genetic diversity of *Actinobacillus lignieresii* isolates from different hosts. *Int J Acta Vet Scand* 53(1): 6. http://www.actavetscand.com/content/53/1/6

AEROMONAS SPP.

Species of *Aeromonas* are Gram-negative, non-spore-forming, rod-shaped, facultatively anaerobic bacteria that occur ubiquitously and autochthonously in aquatic environments. *Aeromonas* spp. are Gram-negative rods of the family Vibrionaceae. They are normal water inhabitants and are part of the regular flora of poikilotherm, and homeotherm animals. They can be isolated from many foodstuffs (green vegetables, raw milk, ice cream, meat, and seafood).

While traveler's diarrhea is the most common health problem of international visitors, the major human diseases caused by *Aeromonas* spp. can be classified in two principal groups: septicemia (mainly by strains of *A. veronii* subsp. *sobria* and *A. hydrophila*) and gastroenteritis (any mesophilic *Aeromonas* spp. but principally *A. hydrophila* and *A. veronii*). Most epidemiological studies have shown *Aeromonas* spp. in stools were more often associated with diarrhea than with the carrier state; an association with the consumption of untreated water was also conspicuous. Acute self-limited diarrhea is more frequent in young children and in older patients. Chronic enterocolitis may also be observed. Fever, vomiting, and fecal leukocytes or erythrocytes (colitis) may be present. The main putative virulence factors are exotoxins and endotoxin (lipopolysaccharide, LPS).

SOURCES

Merino S, Rubitre X, Knochel S, and Tomas JM. 1995. Emerging pathogens: *Aeromonas* spp. *Int J Food Microbiol* 28(2): 157–168.

Jordi Vila, Laboratori de Microbiologia, Institut d'Infeccions i Immunologia, Institut d'Investigacions Biomèdiques August Pi i Sunyer, Fascultat de Medicina, Universitat de Barcelona, Villarroel, 170; 08036 Barcelona, Spain; fax: 34.93.2279372. Opportunisitic pathogen that can cause diarrheal diseases in otherwise healthy individuals and can cause wound infections. *Int J Food Microbiol* 28: 2157–2168.

ASCARIS LUMBRICOIDES

An estimated 807–1,221 million people in the world are infected with *Ascaris lumbricoides* (sometimes called just *Ascaris*). *Ascaris*, hookworm, and whipworm are known as soil-transmitted helminths (parasitic worms). Together, they account for a major burden of disease worldwide. Ascariasis is now uncommon in the United States, according to the Centers for Disease Control and Prevention.

Ascaris lives in the intestine and *Ascaris* eggs are passed in the feces of infected persons. If the infected person defecates outside (near bushes, in a garden, or field) or if the feces of an infected person are used as fertilizer, eggs are deposited on soil. They can then mature into a form that is infective. Ascariasis is caused by ingesting eggs. This can happen when hands or fingers that have contaminated dirt on them are put in the mouth or by consuming vegetables or fruits that have not been carefully cooked, washed, or peeled.

People infected with *Ascaris* often show no symptoms. If symptoms do occur, they can be light and include abdominal discomfort. Heavy infections can cause intestinal blockage and impair growth in children. Other symptoms such as cough are due to migration of the worms through the body. Ascariasis is treatable with medication prescribed by your health care provider.

Ascaris lumbricoides is a nematode (roundworm) that inhabits the intestines of humans. It measures 13–35 cm in length and may live in the gut for 6–24 months. Infection is normally from food contaminated by soil containing feces from the worm. It is prevalent in developing countries, where there is often a combination of poor sanitation and a host made vulnerable by malnutrition, iron-deficiency anemia, or impairment of growth.

If only a few worms are present, there may be no symptoms initially, but during the migratory phase, the larvae may penetrate into the tissues and circulate around the body via the blood and lymphatic systems, commonly to the lungs. In the lungs, the larvae penetrate the pulmonary capillaries to enter the alveoli, from where they ascend into the throat and descend back into the gut, where they may grow as large as 35 cm in length.

SOURCES

Centers for Disease Control and Prevention, 1600 Clifton Rd., Atlanta, GA 30333. www.cdc.gov/parasites/ascariasis 2013

Dora-Laskey A et al. 2012. *Ascaris Lumbricoides*-Medscape Reference. http://emedicine.medscape.com/article/788398-overview

Patient.co.uk. http://www.patient.co.uk/doctor/ascaris-lumbricoides

BACILLACEAE SPP.

See Bacillus spp.

Members of the family Bacillaceae produce endospores; most are Gram positive, motile by lateral or peritrichous flagella (having flagella over the entire surface) or nonmotile, and aerobic, facultative, or anaerobic.

The genus *Bacillus, of the family Bacillaceae*, consists of a heterogeneous group of Gram-positive, heterotrophic, aerobic or facultative anaerobic bacilli with the ability to form environmentally resistant, metabolically inert spores. These soil-borne organisms are ubiquitous throughout the world and occupy surprisingly diverse environments. Within this large genus, the *B. cereus* sensu lato group consists of six species, based on classical microbial taxonomy: *B. anthracis (Ba), B. cereus (Bc)*,

B. mycoides, *B. pseudomycoides*, *B. thuringiensis* (*Bt*), and *B. weihenstephanensis*. However, newer molecular phylogenies and comparative genome sequencing suggests that these organisms should be classified as a single species. On the surface, this conclusion seems difficult to reconcile with the varied biological characteristics of these organisms. Some *Bc* strains are thermophiles, while *B. weihenstephanensis* is psychrophilic. By contrast, many members of this group are mesophiles and can be found in a variety of locales, including soil, on plant surfaces, and in the mammalian gastrointestinal microflora. Some members of this group appear to be nonpathogenic, while others cause diverse diseases, including gastroenteritis, food poisoning, endophthalmitis, tissue abscesses, and anthrax. *Bt* strains have the capacity to cause disease in insects and possibly nematodes, while some evidence suggests that *Bc* strains are part of the normal insect gut flora. Nevertheless, whole-genome comparisons between these organisms reveal a surprising similarity in gene content, and Han et al. (1989) have concluded "that differential regulation [of gene content] modulates virulence rather than simple acquisition of virulence factor genes," a conclusion confirmed by other studies. Consequently, we will refer to these organisms as the *Bc* species-group, to reflect the extremely close phylogenetic relationships between these organisms.

SOURCES

Han K et al. 1989. Synergistic activation and repression of transcription by Drosophila homeobox proteins. *Cell* 56(4): 573–583.
http://www.biomedcentral.com/1471-2164/12/430
Schmidt T, Scott EJ, and Dyer DW. 2011. Whole-genome phylogenies of the family Bacillaceae and expansion of the sigma factor gene family in the *Bacillus cereus* species-group. *BMC Genomics* 12: 430. doi:10.1186/1471-2164-12-430. http://www.biomedcentral.com/1471-2164/12/430

BACILLUS SPP.

Bacillus cereus is a Gram-positive, facultatively anaerobic, endospore-forming, large rod. These and other characteristics, including biochemical tests, are used to differentiate and confirm the presence of *B. cereus*, although these characteristics are shared with *B. mycoides*, *B. pseudomycoides*, *B. thuringiensis*, and *B. anthracis*. Differentiation of these organisms depends on:

determination of motility (most *B. cereus* strains are motile)
presence of toxin crystals (*B. thuringiensis*)
hemolytic activity (*B. cereus* and others are beta-hemolytic, whereas *B. anthracis* usually is nonhemolytic)
rhizoid growth, which is characteristic of *B. cereus* var. *mycoides*

Bacilli are an extremely diverse group of bacteria that include both the causative agent of anthrax (*Bacillus anthracis*) as well as several species that synthesize important antibiotics. In addition to medical uses, bacillus spores, due to their extreme tolerance to both heat and disinfectants, are used to test heat sterilization techniques and chemical disinfectants. *Bacillus* is a genus of Gram-positive, rod-shaped (bacillus), bacteria and a member of the phylum Firmicutes. *Bacillus* species can be obligate aerobes (oxygen reliant) or facultative anaerobes (having the ability to be aerobic or anaerobic). They will test positive for the enzyme catalase when there has been oxygen used or present. Ubiquitous in nature, *Bacillus* includes both free-living (nonparasitic) and parasitic pathogenic species. Under stressful environmental conditions, the bacteria can produce oval endospores that are not true spores but which the bacteria can reduce themselves to and remain in a dormant state for very long periods. These characteristics originally defined the genus, but not all such species are closely related, and many have been moved to other genera of Firmicutes.

Bacilli cause an array of infections, from ear infections to meningitis and urinary tract infections to septicemia. Mostly, they occur as secondary infections in immunodeficient hosts or otherwise-compromised hosts. They may exacerbate previous infections by producing tissue-damaging toxins or metabolites that interfere with treatment.

The most well-known disease caused by bacilli is anthrax, caused by *Bacillus anthracis*. Anthrax has a long history with humans. It has been suggested that the fifth and sixth plagues of Egypt recorded in the Bible (the fifth attacking animals, the sixth, known as the plague of the boils, attacking humans) are descriptive of *B. anthracis*. Bacilli are an extremely diverse group of bacteria that include both the causative agent of anthrax (*Bacillus anthracis*) as well as several species that synthesize important antibiotics. In addition to medical uses, bacillus spores, due to their extreme tolerance to both heat and disinfectants, are used to test heat sterilization techniques and chemical disinfectants. Anthrax has more recently been brought to our attention as a possible method for bioterrorism. The recent anthrax mailings have brought acute public attention to the issue and sparked extensive research into the devastating disease.

Anthrax is primarily a disease of herbivores, who acquire the bacterium by eating plants with dust that contains anthrax spores. Humans contract the disease in three different ways. Cutaneous anthrax occurs when a human comes into contact with the spores from dust particles or a contaminated animal or carcass through a cut or abrasion. Cutaneous anthrax accounts for 95% of anthrax cases worldwide. During a 2- to 3-day incubation period, the spores germinate, vegetative cells multiply, and a papule develops. Over the following days, the papule ulcerates, dries, and blackens to form the characteristic eschar. The process is painless unless infected with another pathogen.

Gastrointestinal anthrax is contracted by ingesting contaminated meat. It occurs in the intestinal mucosa when the organisms invade the mucosa through a preexisting lesion. It progresses the same way as cutaneous anthrax. Although it is extremely rare in developed countries, it has a very high mortality rate.

Pulmonary anthrax is the result of inhaled spores that are transported to the lymph nodes, where they germinate and multiply. They are then taken into the bloodstream and lymphatics, culminating in systemic arthritis, which is usually fatal.

SOURCES

Turnbull PCB. 1996. *Bacillus*. In *Barons Medical Microbiology* 4th edition. Galveston, TX: University of Texas Medical Branch. http://en.wikipedia.org/wiki/Bacillus
Bacillus. Microbe Wiki. 2010. http://microbewiki.kenyon.edu/index.php/Bacillus

Turnbull PCB. *Bacillus*. In *Medical Microbiology NCBI Bookshelf* 4th edition, chapter 15. http://www.ncbi.nlm.nih.gov/books/NBK7699

Food and Drug Administration. 2012. *Bad Bug Book. Foodborne Pathogenic Microorganisms and Natural Toxins*. 2nd edition. Washington, DC: Food and Drug Administration.

BORDETELLA SPP.

Bordetella spp. are considered opportunistic pathogens. Thus, they predominantly infect immunosuppressed and immunocompromised individuals. Several *Bordetella* species have been associated with respiratory disease in humans. *Bordetella pertussis* still poses an important health threat in developing countries. Although *B. avium* is thought to be strictly an animal pathogen that causes tracheobronchitis in wild and domesticated birds, infections in birds share many of the clinical and histopathologic features of disease in mammals caused by *B. pertussis* and *B. bronchiseptica*. Human cases of respiratory disease associated with B. avium have only recently been reported in patients with cystic fibrosis.

Pertussis or whooping cough is a highly infectious respiratory disease caused by *Bordetella pertussis*. In vaccinating countries, infants, adolescents, and adults are relevant patient groups.

Whooping cough is a major cause of infant and childhood mortality. The World Health Organization (WHO) reported about 16 million pertussis cases worldwide in 2008, with 95% of cases occurring in developing countries, and more than 100,000 children died from this disease. Pertussis remains endemic despite the introduction of a vaccination program in 1974. During 2003–2007, there were 43,482 cases or an incidence of 4.1 per 100,000 people reported from 20 European countries. In the United States, the incidence of pertussis also increased from 3.53 in year 2007 to 5.54 per 100,000 in year 2009. Although pertussis is always classified as a disease of infants and children, an increasing number of cases in the adolescent and adult groups were also observed.

Sources

Emerging Infectious Diseases 15(1), January 2009.

Jusot V, Aberrane S, Ale F, Laouali B, Moussa I, Alio SA, Adehossi E, Collard JM, and Grais RF. 2014. Prevalence of *Bordetella* infection in a hospital setting in Niamey, Niger. *J Trop Pediatr* 60(3): 223–230.

Ting TX, Hashim R, Ahmad N, and Abdullah KH. 2013. Detection of *Bordetella pertussis* from clinical samples by culture and end-point PCR in Malaysian patients. *Int J Bacteriol* 2013, Article ID 324136. http://dx.doi.org/10.1155/2013/324136

CAMPYLOBACTER JEJUNI

Campylobacter is one of the most common causes of diarrheal illness in the United States. Most cases occur as isolated, sporadic events, not as part of recognized outbreaks. Active surveillance through the Foodborne Diseases Active Surveillance Network (FoodNet) indicates that about 14 cases are diagnosed each year for each 100,000 persons in the population. Many more cases go undiagnosed or unreported, and campylobacteriosis is estimated to affect over 1.3 million persons every year. Campylobacteriosis occurs much more frequently in the summer months than in the winter. The organism is isolated from infants and young adults more frequently than from persons in other age groups and from males more frequently than females.

It is a non-spore-forming, Gram-negative rod with a curved to S-shaped morphology. Many strains display motility, which is associated with the presence of a flagellum at one or both of the polar ends of this bacterium.

Members of the *Campylobacter* genus are microaerophilic, that is, they grow at lower-than-atmospheric oxygen concentrations. Most grow optimally at oxygen concentrations from 3% to 5%. Thus, these bacteria generally are fairly fragile in the ambient environment and somewhat difficult to culture in the laboratory. Additional conditions to which *C. jejuni* are susceptible include drying, heating, freezing, disinfectants, and acidic conditions.

Other *Campylobacter* species, such as *C. coli* and *C. fetus*, also cause foodborne diseases in humans; however, more than 80% of *Campylobacter* infections are caused by *C. jejuni*. *Campylobacter coli* and *C. jejuni* cause similar disease symptoms. *Campylobacter fetus* infections often are associated with animal contact or consumption of contaminated foods and beverages and are especially problematic for fetuses and neonates, in whom the mortality rate may be up to 70%.

Campylobacteriosis is an infectious disease caused by bacteria of the genus *Campylobacter*. Most people who become ill with campylobacteriosis get diarrhea, cramping, abdominal pain, and fever within 2 to 5 days after exposure to the organism. The diarrhea may be bloody and can be accompanied by nausea and vomiting. The illness typically lasts about 1 week. Some infected persons do not have any symptoms. In persons with compromised immune systems, *Campylobacter* occasionally spreads to the bloodstream and causes a serious life-threatening infection.

It does not commonly cause death. It has been estimated that approximately 76 persons with *Campylobacter* infections die each year.

Campylobacter jejuni is commonly found in animal feces. *Campylobacter* is the most common bacterial cause of diarrhea in the United States; over 2 million cases are reported each year. Although *Campylobacter* does not commonly cause death, it is estimated that approximately 100 persons with *Campylobacter* infections die each year.

In general, the infective dose, the minimum number of ingested *Campylobacter* cells that can cause infection, is thought to be about 10,000. However, in trials, as few as 500 ingested *Campylobacter* cells led to disease, not necessarily death, in volunteers. Differences in infectious dose likely can be attributed to several factors, such as the type of contaminated food consumed and the general health of the exposed person.

The incubation period, from time of exposure to onset of symptoms, generally is 2 to 5 days.

A small percentage of patients develop complications that may be severe. These include bacteremia and infection of various organ systems, such as meningitis, hepatitis, cholecystitis, and pancreatitis. An estimated 1.5 cases of bacteremia occur for every 1,000 cases of gastroenteritis. Infections also may lead, although rarely, to miscarriage or neonatal sepsis.

Autoimmune disorders are another potential long-term complication associated with campylobacteriosis, for example, Guillain-Barré syndrome (GBS). One case of GBS is estimated to develop per 2,000 *C. jejuni* infections, typically 2 to 3 weeks postinfection. Not all cases of GBS appear to be associated with campylobacteriosis, but it is the factor most commonly identified prior to development of GBS. Various studies have shown that up to 40% of GBS patients first had a *Campylobacter* infection. It is believed that antigens present on *C. jejuni* are similar to those in certain nervous tissues in humans, leading to the autoimmune reaction. Reactive arthritis is another potential long-term autoimmune complication. It can be triggered by various kinds of infections and occurs in about 2% of *C. jejuni* gastroenteritis cases.

Hemolytic uremic syndrome and recurrent colitis following *C. jejuni* infection also have been documented.

The major symptoms are fever, diarrhea, abdominal cramps, and vomiting. The stool may be watery or sticky and may contain blood (sometimes occult, i.e., not discernible to the naked eye) and fecal leukocytes (white cells). Other symptoms often present include abdominal pain, nausea, headache, and muscle pain.

Most cases of campylobacteriosis are self-limiting. The disease typically lasts from 2 to 10 days.

Sources

http://www.cdc.gov/pulsenet/pathogens/campylobacter.html
Centers for Disease Control and Prevention (CDC).
Division of Foodborne, Waterborne, and Environmental Diseases (DFWED).
Food and Drug Administration. 2012. *Bad Bug Book. Foodborne Pathogenic Microorganisms and Natural Toxins*. 2nd ed. Washington, DC: Food and Drug Administration.
National Center for Emerging and Zoonotic Infectious Diseases (NCEZID).

CELLULOMONAS

Cellulomonas is a genus of Gram-positive rod-shaped bacteria. One of its main distinguishing features is the ability to degrade cellulose using enzymes such as endoglucanase and exoglucanase. The species are members of the Actinobacteria group.

Cellulomonas spp. are often believed to be of low virulence. There are only a few reports of human infections.

Sources

Glazer AN, and Nikaido H. 2007. *Microbial Biotechnology*. 2nd ed. Cambridge, UK: Cambridge University Press.
Madigan MT, Martinko JM, Dunlap PV, and Clark DP. 2009. *Brock Biology of Microorganisms*. 12th ed. San Francisco: Pearson.

ENTAMOEBA HISTOLYTICA

Entamoeba histolytica is a single-cell anaerobic protozoan parasite responsible for a disease called amoebiasis. It occurs usually in the large intestine and causes internal inflammation, as its name suggests (*histo* = tissue, *lytic* = destroying). Worldwide, 50 million people are infected, mostly in tropical countries in areas of poor sanitation. In industrialized countries, most of the infected patients are immigrants, institutionalized people, and those who have recently visited developing countries.

Inside humans, *Entamoeba histolytica* lives and multiplies as a trophozoite. Trophozoites are oblong and about 15–20 μm in length. To infect other humans, they encyst and exit the body. The life cycle of *Entamoeba histolytica* does not require any intermediate host. Mature cysts (spherical, 12–15 μm in diameter) are passed in the feces of an infected human. Another human can become infected by ingesting them in fecally contaminated water, food, or hands. If the cysts survive the acidic stomach, they transform back into trophozoites in the small intestine. Trophozoites migrate to the large intestine, where they live and multiply by binary fission. Both cysts and trophozoites are sometimes present in the feces. Cysts are usually found in firm stool, whereas trophozoites are found in loose stool. Only cysts can survive longer periods (up to many weeks outside the host) and infect other humans. If trophozoites are ingested, they are killed by the gastric acid of the stomach. Occasionally, trophozoites might be transmitted during sexual intercourse.

Severe infections inflame the mucosa of the large intestine, causing amoebic dysentery. The parasites can also penetrate the intestinal wall and travel to organs such as the liver via the bloodstream, causing extraintestinal amoebiasis. Symptoms of these more severe infections include

- anemia
- appendicitis (inflammation of the appendix)
- bloody diarrhea
- fatigue
- fever
- gas (flatulence)
- genital and skin lesions
- intermittent constipation
- liver abscesses (can lead to death if not treated)
- malnutrition
- painful defecation (passage of the stool)
- peritonitis (inflammation of the peritoneum, which is the thin membrane that lines the abdominal wall)
- pleuropulmonary abscesses
- stomachache
- stomach cramping
- toxic megacolon (dilated colon)
- weight loss

Sources

Food and Drug Administration. 2012. *Bad Bug Book. Foodborne Pathogenic Microorganisms and Natural Toxins*. 2nd ed. Washington, DC: Food and Drug Administration.
http://www.parasitesinhumans.org/entamoeba-histolytica-amoebiasis.html

ENTEROBACTERIA

Enterobacteriaceae is a large family of Gram-negative bacteria that includes, along with many harmless symbionts, many of the more familiar pathogens, such as *Salmonella*, *Escherichia coli*, *Yersinia pestis*, *Klebsiella*, and *Shigella*. Other disease-causing bacteria in this family include *Proteus*, *Enterobacter*, *Serratia*, and *Citrobacter*.

Note: Individual pathogens of this family are discussed separately.

ESCHERICHIA COLI

Escherichia coli (*E. coli*) bacteria normally live in the intestines of people and animals. Most *E. coli* are harmless and actually are an important part of a healthy human intestinal tract. However, some *E. coli* are pathogenic, meaning they can cause illness, either diarrhea or illness outside the intestinal tract. The types of *E. coli* that can cause diarrhea can be transmitted through contaminated water or food or through contact with animals or persons.

Currently, there are four recognized classes of enterovirulent *E. coli* (collectively referred to as the EEC group) that cause gastroenteritis in humans. Among these is the enterohemorrhagic (EHEC) strain designated *E. coli* O157:H7. *E. coli* is a normal inhabitant of the intestines of all animals, including humans. When aerobic culture methods are used, *E. coli* is the dominant species found in feces. Normally, *E. coli* serves a useful function in the body by suppressing the growth of harmful bacterial species and by synthesizing appreciable amounts of vitamins. A minority of *E. coli* strains are capable of causing human illness by several different mechanisms. *E. coli* serotype O157:H7 is a rare variety of *E. coli* that produces large quantities of one or more related, potent toxins that cause severe damage to the lining of the intestine. These toxins (verotoxin [VT], shiga-like toxin] are closely related or identical to the toxin produced by *Shigella dysenteriae*.

E. coli consists of a diverse group of bacteria. Pathogenic *E. coli* strains are categorized into pathotypes. Six pathotypes are associated with diarrhea and collectively are referred to as diarrheagenic *E. coli*.

- Shiga toxin-producing *E. coli* (STEC): STEC may also be referred to as verocytotoxin-producing *E. coli* (VTEC) or EHEC. This pathotype is the one most commonly heard about in the news in association with foodborne outbreaks.
- Enterotoxigenic *E. coli* (ETEC)
- Enteropathogenic *E. coli* (EPEC)
- Enteroaggregative *E. coli* (EAEC)
- Enteroinvasive *E. coli* (EIEC)
- Diffusely adherent *E. coli* (DAEC)

Some kinds of *E. coli* cause disease by making a toxin called Shiga toxin. The bacteria that make these toxins are called Shiga toxin-producing *E. coli*, or STEC for short. You might hear these bacteria called verocytotoxic *E. coli* (VTEC) or enterohemorrhagic *E. coli* (EHEC); these all refer generally to the same group of bacteria. The strain of Shiga toxin-producing *E. coli* O104:H4 that caused a large outbreak in Europe in 2011 was frequently referred to as EHEC. The most commonly identified STEC in North America is *E. coli* O157:H7 (often shortened to *E. coli* O157 or even just O157). When you hear news reports about outbreaks of "*E. coli*" infections, they are usually talking about *E. coli* O157.

In addition to *E. coli* O157, many other kinds (called serogroups) of STEC cause disease. Other *E. coli* serogroups in the STEC group, including *E. coli* O145, are sometimes called "non-O157 STECs." Currently, there are limited public health surveillance data on the occurrence of non-O157 STECs, including STEC O145; many STEC O145 infections may go undiagnosed or unreported.

Compared with STEC O157 infections, identification of non-O157 STEC infections is more complex. First, clinical laboratories must test stool samples for the presence of Shiga toxins. Then, the positive samples must be sent to public health laboratories to look for non-O157 STEC. Clinical laboratories typically cannot identify non-O157 STEC. Other non-O157 STEC serogroups that often cause illness in people in the United States include O26, O111, and O103. Some types of STEC frequently cause severe disease, including bloody diarrhea and hemolytic uremic syndrome (HUS), which is a type of kidney disease.

Sources

Centers for Disease Control and Prevention. http://www.cdc.gov/ecoli/general/index.html?s_cid=cs_002

Food and Drug Administration. 2012. *Bad Bug Book. Foodborne Pathogenic Microorganisms and Natural Toxins*. 2nd ed. Washington, DC: Food and Drug Administration.

KLEBSIELLA PNEUMONIA

Klebsiella pneumoniae is a Gram-negative, nonmotile, encapsulated, lactose-fermenting, facultative anaerobic, rod-shaped bacterium. Although found in the normal flora of the mouth, skin, and intestines, it can cause destructive changes to human lungs if aspirated. In the clinical setting, it is the most significant member of the *Klebsiella* genus of Enterobacteriaceae. *Klebsiella oxytoca* and *Klebsiella rhinoscleromatis* have also been demonstrated in human clinical specimens. In recent years, klebsiellae have become important pathogens in nosocomial infections.

It naturally occurs in the soil, and about 30% of strains can fix nitrogen in anaerobic conditions [2]. As a free-living diazotroph, its nitrogen fixation system has been much studied and is of agricultural interest, as *K. pneumoniae* has been demonstrated to increase crop yields in agricultural conditions.

K. pneumoniae can cause destructive changes to human lungs via inflammation and hemorrhage, with cell death (necrosis) that sometimes produces a thick, bloody, mucoid sputum (currant jelly sputum). These bacteria gain access typically after a person aspirates colonizing oropharyngeal microbes into the lower respiratory tract.

As a general rule, *Klebsiella* infections are seen mostly in people with a weakened immune system. Most often, illness affects middle-aged and older men with debilitating diseases. This patient population is believed to have impaired respiratory host defenses, including persons with diabetes, alcoholism, malignancy, liver disease, chronic obstructive pulmonary diseases (COPDs), glucocorticoid therapy, renal failure, and certain occupational exposures (such as paper mill workers). Many of these infections are obtained when a person is in the hospital (a nosocomial infection) for some other reason. Feces are the most significant source of patient infection, followed by contact with contaminated instruments.

The most common infection caused by *Klebsiella* bacteria outside the hospital is pneumonia, typically in the form of bronchopneumonia and bronchitis. These patients have an increased tendency to develop lung abscess, cavitation, empyema, and pleural adhesions. It has a high death rate of about 50%, even with antimicrobial therapy. The mortality rate can be nearly 100% for persons with alcoholism and bacteremia.

In addition, *K. pneumoniae* has long been recognized as a possible cause of community-acquired pneumonia. Over the past two decades, *K. pneumoniae* has been an exceedingly rare cause of community-acquired pneumonia in North America, Europe, and Australia. Yet, it remains an important cause of severe community-acquired pneumonia in Asia and Africa. In these regions, patients also have the classic risk factor of alcoholism.

Sources

http://en.wikipedia.org/wiki/Klebsiella_pneumoniae

Yu VL, Hansen DS, Ko WC, Sagnimeni A, Klugman KP, von Gottberg A, et al. 2007. Virulence characteristics of *Klebsiella* and clinical manifestations of *K. pneumoniae* bloodstream infections. *Emerg Infect Dis* http://wwwnc.cdc.gov/eid/article/13/7/07–0187.htm.

MICROCOCCUS

Micrococcus is a genus of spherical bacteria in the family Micrococcaceae that is widely disseminated in nature. Micrococci are microbiologically characterized as Gram-positive cocci, 0.5 to 3.5 μm in diameter.

Micrococci are usually not pathogenic. They are normal inhabitants of the human body and may even be essential in keeping the balance among the various microbial flora of the skin. Some species are found in the dust of the air (*M. roseus*), in soil (*M. denitrificans*), in marine waters (*M. colpogenes*), and on the skin or in skin glands or skin-gland secretions of vertebrates (*M. flavus*). Those species found in milk, such as *M. luteus, M. varians,* and *M. freudenreichii,* are sometimes referred to as milk micrococci and can result in spoilage of milk production.

Micrococci have been isolated from human skin, animal and dairy products, and beer. They are found in many other places in the environment, including water, dust, and soil. Micrococcus luteus on human skin transforms compounds in sweat into compounds with an unpleasant odor. Micrococci can grow well in environments with little water or high-salt concentrations. Most are mesophiles; some, like *Micrococcus antarcticus* (found in Antarctica), are psychrophiles.

Though not a spore former, *Micrococcus* cells can survive for an extended period of time. It has been indicated that the survival time can be thousands of years.

Micrococcus is generally thought to be a saprotrophic or commensal organism, although it can be an opportunistic pathogen, particularly in hosts with compromised immune systems, such as patients with human immunodeficiency virus (HIV). It can be difficult to identify *Micrococcus* as the cause of an infection since the organism is normally present in skin microflora, and the genus is seldom linked to disease. In rare cases, death of immunocompromised patients has occurred from pulmonary infections caused by *Micrococcus*. Micrococci may be involved in other infections, including recurrent bacteremia, septic shock, septic arthritis, endocarditis, meningitis, and cavitating pneumonia (immunosuppressed patients).

Sources

Encyclopedia Britannica. http://www.britannica.com/EBchecked/topic/380299/Micrococcus.
http://en.wikipedia.org/wiki/Micrococcus

MYCOBACTERIUM SPP.

Mycobacteria are widespread in nature, but the primary sources are water, soil, cows, and gastrointestinal tracts of animals. *Mycobacterium bovis* is pathogenic for cattle and some other animals but also has been shown to be infectious to humans and therefore is a pathogen of concern to humans. *Mycobacterium* species are considered hardy because of their unique cell walls, which enable them to survive long exposures to chemical disinfectants, including acids, alkalis, and detergents, and because they are able to resist lysis by antibiotics. *Mycobacterium bovis* can survive in the environment for several months in cold, dark, moist conditions and for up to 332 days at a temperature range of 12°C to 24°C.

Mycobacterium bovis, also referred to as *Mycobacterium tuberculosis* var. *bovis*, is a Gram-positive, aerobic, nonmotile, straight or slightly curved, rod-shaped bacterium that lacks an outer cell membrane. It does not have spores or capsules and is classified as an acid-fast bacterium because in staining procedures its lipid-rich cell wall resists decolorization by acids.

Some other species of the genus *Mycobacterium* include *M. tuberculosis*, *M. leprae*, *M. africanum*, *M. avium*, and *M. microti*. Members of the *Mycobacterium* tuberculosis complex, which includes *M. tuberculosis* and *M. bovis*, are the causative agents of human and animal tuberculosis. *Mycobacterium bovis* is a causative agent of foodborne human tuberculosis (although it may also be transmitted via the airborne route if it subsequently infects the lungs and results in active disease).

Mycobacterium bovis causes tuberculosis in cattle and is considered a zoonotic disease that also affects humans. Human tuberculosis caused by this organism is now rare in the United States because of milk pasteurization and culling of infected cattle.

Death can result if the infection is left untreated. The Centers for Disease Control and Prevention (CDC) recently reported that an estimated three deaths (mean) are associated with M. bovis annually in the United States.

The infective dose of *Mycobacterium bovis* in cattle could be as low as 1 CFU (6–10 organisms), while the precise infective dose for humans is still unknown; it is suggested to be on the order of tens to millions of organisms.

Symptoms generally appear months to years after initial infection. Some infected persons do not show any signs of the disease.

Ingestion of food contaminated with *M. bovis* can result in infection of the gastrointestinal tract or other parts of the body, for example, the lungs or the lymph nodes. The disease may result in death if untreated.

Typical symptoms include fever, night sweats, fatigue, loss of appetite, and weight loss. Other symptoms depend on the part of the body affected, for example, chronic cough, blood-stained sputum, or chest pain, if the lungs are affected, or diarrhea, abdominal pain, and swelling, if the gastrointestinal tract is affected. Infections in humans also may be asymptomatic.

Duration of illness depends on the immune status of the infected person. Symptoms could last for months or years, which necessitates a longer treatment period. Individuals with symptoms of lung involvement should avoid public settings until told by their health care providers that they are no longer a risk to others.

The route of entry is mostly through ingestion (oral). Inhalation and direct contact with mucous membranes or broken skin, although not common, are also potential routes of exposure.

Regarding the pathway, *M. bovis* can be taken up by alveolar macrophages in the lung, especially if transmission is by the aerosol route (pulmonary tuberculosis). From there, it is carried to the lymph nodes, where the organism can migrate to other organs. *M. bovis* can multiply in these cells and in interstitial spaces, leading to formation of tubercle lesions. Gastrointestinal tuberculosis also causes the associated lymph nodes to form tubercles, although the organism may not spread to other organs.

An example occurred in March 2004 when a US-born boy aged 15 months in New York City died of peritoneal tuberculosis caused by *Mycobacterium bovis* infection. *M. bovis*, a bacterial species of the *M. tuberculosis* complex, is a pathogen that primarily infects cattle. However, humans also can become infected, most commonly through consumption of unpasteurized milk products from infected cows. In industrialized nations, human tuberculosis caused by *M. bovis* is rare because of milk pasteurization and culling of infected cattle herds. This report summarizes an ongoing, multiagency investigation that has identified 35 cases of human *M. bovis* infection in New York City. Preliminary findings indicate that fresh cheese (e.g., queso fresco) brought to New York City from Mexico was a likely source of infection. No evidence of human-to-human transmission has been found. Products from unpasteurized cow's milk have been associated with certain infectious diseases and carry the risk of transmitting *M. bovis* if imported from countries where the bacterium is common in cattle. All persons should avoid consuming products from unpasteurized cow's milk.

SOURCES

Centers for Disease Control and Prevention. Human tuberculosis caused by *Mycobacterium bovis*—New York City, 2001–2004. 2005. *MMWR Morb Mortal Wkly Rep* 54(24): 605–608.

Food and Drug Administration. 2012. *Bad Bug Book. Foodborne Pathogenic Microorganisms and Natural Toxins*. 2nd ed. Washington, DC: Food and Drug Administration.

NEISSERIA

Neisseria is a large genus of bacteria that colonize the mucosal surfaces of many animals. Of the 11 species that colonize humans, only 2 are pathogens, *N. meningitidis* and *N. gonorrhoeae*. Most gonococcal infections are asymptomatic and self-resolving, and epidemic strains of the meningococcus may be carried in more than 95% of a population where systemic disease occurs at less than 1% prevalence.

Neisseria species are Gram-negative bacteria included among the proteobacteria, a large group of Gram-negative forms.

There are several species of *Neisseria* that are important pathogens. These are discussed separately.

Approximately 2,500 to 3,500 cases of *N. meningitidis* infection occur annually in the United States, with a case rate of about 1 in 100,000. Children younger than 5 years are at greatest risk, followed by teenagers of high school age. Rates in sub-Saharan Africa can be as high as 1 in 1,000 to 1 in 100.

NEISSERIA MENINGITIDIS

Neisseria meningitidis, often referred to as meningococcus, is a bacterium that can cause meningitis and other forms of meningococcal disease, such as meningococcemia, a life-threatening sepsis. *Neisseria meningitidis* is a major cause of illness and death during childhood in industrialized countries and has been responsible for epidemics in Africa and in Asia. The bacteria are round and are often joined in pairs. They are Gram negative since they have outer and inner membranes with a thin layer of peptidoglycan between. Cultures of the bacteria test positive for the enzyme cytochrome c oxidase. It exists as normal flora (nonpathogenic) in the nasopharynx of up to 5%–15% of adults. It causes the only form of bacterial meningitis known to occur epidemically.

Meningococcus is spread through the exchange of saliva and other respiratory secretions during activities like coughing, sneezing, kissing, and chewing on toys. It infects the host cell by sticking to it, mainly with long thin extensions called pili and the surface-exposed proteins Opa and Opc. Although it initially produces general symptoms like fatigue, it can rapidly progress from fever, headache, and neck stiffness to coma and death. The symptoms of meningitis are easily confused with those caused by other organisms, such as *Hemophilus influenzae* and *Streptococcus pneumoniae*. Death occurs in approximately 10% of cases [5]. Those with impaired immunity may be at particular risk of meningococcus (e.g., those with nephrotic syndrome or splenectomy; vaccines are given in cases of removed or nonfunctioning spleens).

NEISSERIA GONORRHOEAE

Neisseria gonorrhoeae infections are acquired by sexual contact and usually affect the mucous membranes of the urethra in males and the endocervix and urethra in females, although the infection may disseminate to a variety of tissues. The pathogenic mechanism involves the attachment of the bacterium to nonciliated epithelial cells via pili and the production of lipopolysaccharide endotoxin. Similarly, the lipopolysaccharide of *Neisseria meningitidis* is highly toxic, and it has an additional virulence factor in the form of its antiphagocytic capsule. Both pathogens produce immunoglobulin A proteases, which promote virulence. Many normal individuals may harbor *Neisseria meningitidis* in the upper respiratory tract, but *Neisseria gonorrhoeae* is never part of the normal flora and is only found after sexual contact with an infected person (or direct contact, in the case of infections in the newborn).

Infection of the genitals can result in a purulent (or pus-like) discharge from the genitals, which may be foul smelling. Symptoms may include inflammation, redness, swelling, and dysuria.

Neisseria gonorrhoeae can also cause conjunctivitis, pharyngitis, proctitis or urethritis, prostatitis, and orchitis.

Conjunctivitis is common in neonates (newborns), and silver nitrate or antibiotics are often applied to their eyes as a preventive measure against gonorrhoea. Neonatal gonorrheal conjunctivitis is contracted when the infant is exposed to *N. gonorrhoeae* in the birth canal and can lead to corneal scarring or perforation, resulting in blindness in the neonate.

Disseminated *N. gonorrhoeae* infections can occur, resulting in endocarditis, meningitis, or gonococcal dermatitis-arthritis syndrome. Dermatitis-arthritis syndrome presents with arthralgia, tenosynovitis, and painless nonpruritic (nonitchy) dermatitis.

Infection of the genitals in females with *N. gonorrhoeae* can result in pelvic inflammatory disease if left untreated, which can result in infertility. Pelvic inflammatory disease results if *N. gonorrhoeae* travels into the pelvic peritoneum (via the cervix, endometrium, and fallopian tubes). Infertility is caused by inflammation and scarring of the fallopian tube. Infertility is a risk to 10% to 20% of the females infected with *N. gonorrhoeae*.

The prevalence of gonorrhea (*Neisseria gonorrheae*) in the United States and abroad, especially in underdeveloped and developing countries, has decreased in the last two decades. Recently, though, higher rates of infection have been reported due to the increase of antimicrobial-resistant gonococci.

Sources

http://biology.kenyon.edu/slonc/bio38/stancikd_02/What_is_Neisseria_gonorrhoeae.html.
http://en.wikipedia.org/wiki/Neisseria_meningitidis#Epidemiology.
Todar's Online Textbook. http://textbookofbacteriology.net/neisseria.html.

PROTEUS SPP.

Proteus is a genus of Gram-negative proteobacteria. *Proteus* bacilli are widely distributed in nature as saprophytes, being found in decomposing animal matter, in sewage, in manure soil, and in human and animal feces. They are opportunistic pathogens, commonly responsible for urinary and septic infections, often nosocomially.

Three species—*P. vulgaris*, *P. mirabilis*, and *P. penneri*—are opportunistic human pathogens. *Proteus* includes pathogens responsible for many human urinary tract infections. *P. mirabilis* causes wound and urinary tract infections. Most strains of *P. mirabilis* are sensitive to ampicillin and cephalosporins. *P. vulgaris* is not sensitive to these antibiotics. However, this organism is isolated less often in the laboratory and usually only targets immunosuppressed individuals. *P. vulgaris* occurs naturally in the intestines of humans and a wide variety of animals and in manure, soil, and polluted waters. *P. mirabilis*, once attached to the urinary tract, infects the kidney more commonly than *Escherichia coli*. *P. mirabilis* is often found as a free-living organism in soil and water.

About 10%–15% of kidney stones are struvite stones, caused by alkalization of the urine by the action of the urease enzyme (which splits urea into ammonia and carbon dioxide) of *Proteus* (and other) bacterial species.

Sources

http://en.wikipedia.org/wiki/Proteus_(bacterium)

PSEUDOMONAS SPP.

Pseudomonas infections are caused by any of several types of the Gram-negative *Pseudomonas* bacteria, especially *Pseudomonas aeruginosa*. Infections range from mild external ones (affecting the ear or hair follicles) to serious internal infections (affecting the lungs, bloodstream, or heart).

Pseudomonas bacteria, including *Pseudomonas aeruginosa*, are present throughout the world in soil and water. These bacteria favor moist areas, such as sinks, toilets, inadequately chlorinated swimming pools and hot tubs, and outdated or inactivated antiseptic solutions. These bacteria may temporarily reside in the skin, ears, and intestine of healthy people.

Pseudomonas aeruginosa infections range from minor external infections to serious, life-threatening disorders. Infections occur more often and tend to be more severe in people who

- Are weakened (debilitated) by certain severe disorders
- Have diabetes or cystic fibrosis
- Are hospitalized
- Have a disorder that weakens the immune system, such as human immunodeficiency virus (HIV) infection
- Take drugs that suppress the immune system, such as those used to treat cancer or to prevent rejection of transplanted organs

These bacteria can infect the blood, skin, bones, ears, eyes, urinary tract, heart valves, and lungs, as well as wounds (such as burns, injuries, or wounds made during surgery). Use of medical devices, such as catheters inserted into the bladder or a vein, breathing tubes, and mechanical ventilators, increases the risk of *Pseudomonas aeruginosa* infections. These infections are commonly acquired in hospitals.

P. aeruginosa is a rod-shaped organism that can be found in soil, water, plants, and animals. Because it rarely causes disease in healthy persons but infects those who are already sick or who have weakened immune systems, it is called an opportunistic pathogen. Opportunistic pathogens are organisms that do not ordinarily cause disease but multiply freely in persons whose immune systems are weakened by illness or medication. Such persons are said to be immunocompromised. Patients with acquired immunodeficiency syndrome (AIDS) have an increased risk of developing serious *Pseudomonas* infections. Hospitalized patients are another high-risk group because *P. aeruginosa* is often found in hospitals. Infections that can be acquired in the hospital are sometimes called nosocomial diseases.

- Heart and blood. *P. aeruginosa* is the fourth-most-common cause of bacterial infections of the blood (bacteremia). Bacteremia is common in patients with blood cancer and patients who have *Pseudomonas* infections elsewhere in the body. *P. aeruginosa* infects the heart valves of intravenous drug abusers and persons with artificial heart valves.
- Bones and joints. *Pseudomonas* infections in these parts of the body can result from injury, the spread of infection from other body tissues, or bacteremia. Persons at risk for *Pseudomonas* infections of the bones and joints include diabetics, intravenous drug abusers, and bone surgery patients.
- Central nervous system. *P. aeruginosa* can cause inflammation of the tissues covering the brain and spinal cord (meningitis) and brain abscesses. These infections may result from brain injury or surgery, the spread of infection from other parts of the body, or bacteremia.
- Eye and ear. *P. aeruginosa* can cause infections in the external ear canal—so-called swimmer's ear—that usually disappear without treatment. The bacterium can cause a more serious ear infection in elderly patients, possibly leading to hearing problems, facial paralysis, or even death. *Pseudomonas* infections of the eye usually follow an injury. They can cause ulcers of the cornea that may cause rapid tissue destruction and eventual blindness. The risk factors for *Pseudomonas* eye infections include wearing soft extended-wear contact lenses; using topical corticosteroid eye medications; being in

Appendix: Details of the Pathogens and Their Diseases 135

a coma; having extensive burns; undergoing treatment in an intensive care unit; and having a tracheostomy or endotracheal tube.
- Urinary tract. Urinary tract infections can be caused by catheterization, medical instruments, and surgery.
- Lung. Risk factors for *P. aeruginosa* pneumonia include cystic fibrosis; chronic lung disease; immunocompromised condition; being on antibiotic therapy or a respirator; and congestive heart failure. Patients with cystic fibrosis often develop *Pseudomonas* infections as children and suffer recurrent attacks of pneumonia.
- Skin and soft tissue. Even healthy persons can develop a *Pseudomonas* skin rash following exposure to the bacterium in contaminated hot tubs, water parks, whirlpools, or spas. This skin disorder is called *Pseudomonas* or "hot tub" folliculitis and is often confused with chickenpox. Severe skin infection may occur in patients with *P. aeruginosa* bacteremia. The bacterium is the second-most-common cause of burn wound infections in hospitalized patients.

SOURCES

The Free Dictionary.
http://medical-dictionary.thefreedictionary.com/Pseudomonas+Infections
http://www.merckmanuals.com/home/infections/bacterial_infections/pseudomonas_infections.html

SALMONELLA SPP.

Salmonella is a motile, non-spore-forming, Gram-negative, rod-shaped bacterium in the family Enterobacteriaceae and the tribe Salmonellae. Nonmotile variants include *S. gallinarum* and *S. pullorum*. The genus *Salmonella* is divided into two species that can cause illness in humans:

Salmonella enterica
Salmonella bongori

Salmonella enterica, which is of the greatest public health concern, is comprised of six subspecies:

Salmonella enterica subsp. *enterica* (I)
Salmonella enterica subsp. *salamae* (II)
Salmonella enterica subsp. *arizonae* (IIIa)
Salmonella enterica subsp. *diarizonae* (IIIb)
Salmonella enterica subsp. *houtenae* (IV)
Salmonella enterica subsp. *indica* (VI)

Salmonella is further subdivided into serotypes based on the Kaufmann–White typing scheme first published in 1934, which differentiates *Salmonella* strains by their surface and flagellar antigenic properties. *Salmonella* spp. are commonly referred to by their serotype names. For example, *Salmonella enterica* subsp. *enterica* is further divided into numerous serotypes, including *Salmonella* serotype Enteritidis and *Salmonella* serotype Typhimurium, which are common in the United States. (Note that species names are italicized, but serotype names are not.) When Kaufmann first proposed the scheme, 44 serotypes had been discovered. As of 2007, the number of serotypes discovered was 2,579.

Salmonella spp. are bacteria that cause one of the most common forms of food poisoning worldwide. There are over 2,500 different types of *Salmonella* spp., but they all produce a similar clinical picture to other forms of infective gastroenteritis.

Salmonellosis is an infection with bacteria called *Salmonella*. *Salmonella* germs have been known to cause illness for over 100 years. They were discovered by an American scientist named Salmon, for whom they are named.

Most persons infected with *Salmonella* develop diarrhea, fever, and abdominal cramps 12 to 72 hours after infection. The illness usually lasts 4 to 7 days, and most persons recover without treatment. However, in some persons, the diarrhea may be so severe that the patient needs to be hospitalized. In these patients, the *Salmonella* infection may spread from the intestines to the bloodstream and then to other body sites and can cause death unless the person is treated promptly with antibiotics. The elderly, infants, and those with impaired immune systems are more likely to have a severe illness.

Every year, approximately 42,000 cases of salmonellosis are reported in the United States. Because many milder cases are not diagnosed or reported, the actual number of infections may be 29 or more times greater. There are many different kinds of *Salmonella* bacteria. *Salmonella* serotype Typhimurium and *Salmonella* serotype Enteritidis are the most common in the United States. Salmonellosis is more common in the summer than winter.

Children are the most likely to get salmonellosis. The rate of diagnosed infections in children less than 5 years old is higher than the rate in all other persons. Young children, the elderly, and the immunocompromised are the most likely to have severe infections. It is estimated that approximately 400 persons die each year with acute salmonellosis.

Salmonella causes two kinds of illness:

Gastrointestinal illness causes nausea, vomiting, diarrhea, cramps, and fever, with symptoms generally lasting a couple of days and tapering off within a week. In otherwise-healthy people, the symptoms usually go away by themselves, but long-term arthritis may develop.
Typhoidal illness causes high fever, diarrhea or constipation, aches, headache, lethargy (drowsiness or sluggishness), and sometimes a rash. It is a very serious condition; up to 10% of people who do not receive treatment may die. Many kinds of food can become contaminated with the first type, from meats and eggs to fruits and vegetables, and even dry foods, like spices and raw tree nuts. The typhoidal illness usually is associated with sewage-contaminated drinking water or crops irrigated with sewage-contaminated

water. Some pets, like turtles and other reptiles and chicks, can carry *Salmonella*, which can spread to anything that comes into contact with the pet. For example, a pet owner can, through unwashed hands, contaminate foods or even his or her own face with *Salmonella*. This bacterium is hard to wash off food, even with soapy water, so important measures for preventing foodborne illness from *Salmonella* include thorough cooking, hand washing, keeping raw foods separated from cooked foods, and keeping foods at the correct temperature (refrigerate foods at 40°F or below). In people with weak immune systems, *Salmonella* can spread to other organs and cause very serious illness.

SOURCES

Centers for Disease Control and Prevention (CDC). http://www.cdc.gov/salmonella/general/
Food and Drug Administration. 2012. *Bad Bug Book. Foodborne Pathogenic Microorganisms and Natural Toxins*. 2nd ed. Washington, DC: Food and Drug Administration.
Patient.co.uk. http://www.patient.co.uk/doctor/salmonella-gastroenteritis

SERRATIA PLYMUTHICA

Serratia is a genus of Gram-negative, facultatively anaerobic, rod-shaped bacteria of the Enterobacteriaceae family. The most common species in the genus, *Serratia marcescens*, is normally the only pathogen and usually causes nosocomial infections. However, rare strains of *Serratia plymuthica*, *S. erratia liquefaciens*, *Serratia rubidaea*, and *Serratia odoriferae* have caused diseases through infection.

In the hospital, *Serratia* species tend to colonize the respiratory and urinary tracts, rather than the gastrointestinal tract, in adults.

Serratia infection is responsible for about 2% of nosocomial infections of the bloodstream, lower respiratory tract, urinary tract, surgical wounds, and skin and soft tissues in adult patients. Outbreaks of *S. marcescens* meningitis, wound infections, and arthritis have occurred in pediatric wards.

Serratia infection has caused endocarditis and osteomyelitis in people addicted to heroin.

Cases of *Serratia* arthritis have been reported in outpatients receiving intra-articular injections.

Serratia marcescens is ubiquitous. It is commonly found in soil, water, plants, and animals. It is widely present in nonpotable water in underdeveloped countries due to poor chlorination.

Serratia marcescens is an opportunistic pathogen that causes nosocomial infections. It is resistant to many antibiotics traditionally used to treat bacterial infections, such as penicillin and ampicillin.

SOURCES

http://en.wikipedia.org/wiki/Serratia
https://microbewiki.kenyon.edu/index.php/Serratia_marcescens

STAPHYLOCOCCUS AUREUS

Staphylococcus aureus (staph) is a type of bacteria that about 30% of people carry in their noses. Most of the time, staph does not cause any harm; however, sometimes staph causes infections. In health care settings, these infections can be serious or fatal, including

- Bacteremia or sepsis when bacteria spread to the bloodstream
- Pneumonia, which predominantly affects people with underlying lung disease, including those on mechanical ventilators
- Endocarditis (infection of the heart valves), which can lead to heart failure or stroke
- Osteomyelitis (bone infection), which can be caused by staph bacteria traveling in the bloodstream or put there by direct contact, such as following trauma (puncture wound of foot or intravenous drug abuse)

Staph bacteria can also become resistant to certain antibiotics. These drug-resistant staph infections include those cased by methicillin-resistant *Staphylococcus aureus* (MRSA), vancomycin-intermediate *Staphylococcus aureus* (VISA), and vancomycin-resistant *Staphylococcus aureus* (VRSA).

Staphylococcal species are Gram-positive, nonmotile, catalase-positive, small, spherical bacteria (cocci), which, on microscopic examination, appear in pairs, short chains, or bunched in grape-like clusters. Staphylococci are ubiquitous and impossible to eradicate from the environment. Many of the 32 species and subspecies in the genus *Staphylococcus* are potentially found in foods due to environmental, human, and animal contamination.

Several staphylococcal species, including both coagulase-negative and coagulase-positive strains, have the ability to produce highly heat-stable enterotoxins that cause gastroenteritis in humans. *S. aureus* is the etiologic agent predominantly associated with staphylococcal food poisoning.

S. aureus is a versatile human pathogen capable of causing staphylococcal food poisoning, toxic shock syndrome, pneumonia, postoperative wound infection, and nosocomial bacteremia.

S. aureus produces a variety of extracellular products, many of which act as virulence factors. Staphylococcal enterotoxins can act as superantigens capable of stimulating an elevated percentage of T cells.

S. aureus is one of the most resistant non-spore-forming human pathogens and can survive for extended periods in a dry state. Staphylococci are mesophilic. *S. aureus* growth, in general, ranges from 7°C to 47.8°C, with 35°C the optimum temperature for growth. The growth pH range is between 4.5 and 9.3, with an optimum between 7.0 and 7.5. Staphylococci are atypical in that they are able to grow at low levels of water activity. For the most part, strains of *S. aureus* are highly tolerant to salts and sugars.

This bacterium, often called "staph" for short, can cause food poisoning. It is common in the environment and can be found in soil, water, and air and on everyday objects and surfaces. It can live in humans and animals. *Staphylococcus aureus* is found in foods and can make toxins (enterotoxins) that might not be destroyed by cooking, although the bacterium itself can be destroyed by heat. These toxins can cause nausea, stomach cramps, vomiting, and diarrhea. In more severe cases, the toxins may cause loss of body fluid (dehydration), headache, muscle cramps, and temporary changes in blood pressure and heart rate. The illness usually is intense but normally lasts from just a few hours to a day. The toxins are fast-acting; they cause symptoms within 1 to 7 hours after contaminated food is eaten. Follow basic food safety tips to help protect yourself from this illness. Outbreaks often have been linked to foods that require a lot of handling when they are being processed or prepared or were not kept at proper refrigerator temperature (40°F or below). To help protect yourself, it is especially important to wash your hands well when handling food, properly clean your cooking equipment and surfaces, keep your cooked foods from touching raw foods or unclean equipment or surfaces, and keep foods refrigerated at 40°F or below. Examples of foods that have been linked to this type of food poisoning include meat and meat products; poultry and egg products; salads, such as egg, tuna, chicken, potato, and macaroni; bakery products, such as cream-filled pastries, cream pies, and chocolate éclairs; sandwich fillings; and milk and dairy products. Good hygienic practices are essential.

Sources

Centers for Disease Control and Prevention (CDC). http://www.cdc.gov/hai/organisms/staph.html

Food and Drug Administration. *Bad Bug Book. Foodborne Pathogenic Microorganisms and Natural Toxins*. 2nd ed. Washington, DC: Food and Drug Administration.

STREPTOCOCCUS FAECALIS (ENTEROCOCCUS FAECALIS), *STREPTOCOCCUS PYOGENES*, AND *STREPTOCOCCUS PNEUMONIAE*

Enterococcus faecalis, formerly classified as part of the group D *Streptococcus* system, is a Gram-positive, commensal bacterium inhabiting the gastrointestinal tracts of humans and other mammals. Like other species in the genus *Enterococcus*, *E. faecalis* can cause life-threatening infections in humans, especially in the nosocomial (hospital) environment, where the naturally high levels of antibiotic resistance found in *E. faecalis* contribute to its pathogenicity.

Streptococcus A is not a major cause of foodborne illness, although serious complications occasionally develop if foodborne illness does occur. Streptococci can be found on the skin; the mucous membranes of the mouth, respiratory, alimentary, and genitourinary tracts of human and animals; and in some plants, soil, and bodies of dirty water. They are opportunistic pathogens. Optimum incubation temperature is usually 37°C, with relatively wide variations among species.

The genus *Streptococcus* is comprised of Gram-positive, catalase-negative, microaerophilic cocci that are nonmotile and occur in chains or pairs, and in long chains in broth culture. Cells are normally spherical, ovoid, and less than 2 μm in diameter.

E. faecalis has frequently been found in root canal-treated teeth in prevalence values ranging from 30% to 90% of the cases. Root canal-treated teeth are about nine times more likely to harbor *E. faecalis* than cases of primary infections.

E. faecalis can cause endocarditis and bacteremia, urinary tract infections, meningitis, and other infections in humans. Several virulence factors are thought to contribute to *E. faecalis* infections. A plasmid-encoded hemolysin, called the cytolysin, is important for pathogenesis in animal models of infection, and the cytolysin in combination with high-level gentamicin resistance is associated with a fivefold increase in risk of death in human bacteremia patients. A plasmid-encoded factor called aggregation substance is also important for virulence in animal models of infection.

In otherwise-healthy people, most cases of foodborne *Streptococcus* infection are relatively mild. In patients who develop invasive disease (most likely to occur in people with underlying health issues, such as those who are immunocompromised), the death rate is estimated at 13%.

Contaminated food is one way you can be infected with *Streptococcus*, the bacterium that causes it. *Streptococcus* is not a leading cause of illness from food, but the illness that it does cause can develop into more serious problems.

Some people infected with foodborne *Streptococcus* have no symptoms, but those who do will start to have them in about 1 to 3 days after eating contaminated food. They may start with red, sore throat (with or without white patches); painful swallowing; high fever; nausea; vomiting; headache; discomfort; and runny nose. The symptoms usually go away in about 4 days. However, 2 or 3 weeks afterward, some people develop scarlet fever, which includes a rash, or rheumatic fever, which can harm the heart and other parts of the body; or, *Streptococcus* could spread to other organs and cause serious illness or death. Children 5 to 15 years old and people with weak immune systems are more likely than others to develop the serious forms of the illness. Infected food handlers are thought to be the main way food is contaminated with *Streptococcus*. In most cases, the food was left at room temperature for too long a time.

The infectious dose for group A *Streptococcus* probably is fewer than 1,000 organisms.

Some foodborne *Streptococcus* group A infections are asymptomatic. Most manifest as pharyngitis (and are commonly referred to as "strep throat"). Although they may be painful and uncomfortable, they usually are relatively mild. However, the infection may also result in complications, such as tonsillitis, scarlet fever, rheumatic fever, and septicemic infections. The symptoms are sore, inflamed throat, on which there are white patches.

Streptococcus pneumoniae is a bacterium commonly found in the nose and throat. The bacterium can sometimes cause severe illness in children, the elderly, and other people with weakened immune systems. *Streptococcus pneumoniae* is the most common cause of middle-ear infections, sepsis (blood infection) in children, and pneumonia in immunocompromised individuals and the elderly. It can also cause meningitis (inflammation of the coverings of the brain and spinal cord) or sinus infections. It is considered invasive when it is found in the blood, spinal fluid, or other normally sterile sites (sites where it is not commonly found).

Appendix: Details of the Pathogens and Their Diseases

Streptococcus pneumoniae causes an acute bacterial infection. The bacterium, also called pneumococcus, was first isolated by Pasteur in 1881 from the saliva of a patient with rabies.

Most *S. pneumoniae* serotypes have been shown to cause serious disease, but only a few serotypes produce the majority of pneumococcal infections. The 10 most common serotypes are estimated to account for about 62% of invasive disease worldwide.

Sources

http://en.wikipedia.org/wiki/Enterococcus_faecalis#Pathogenesis

Food and Drug Administration. *Bad Bug Book. Foodborne Pathogenic Microorganisms and Natural Toxins*. 2nd ed. Washington, DC: Food and Drug Administration.

Iowa Department of Public Health. http://www.idph.state.ia.us/Cade/DiseaseIndex.aspx?disease=Streptococcus%20pneumoniae

Index

Note: Page numbers ending in "f" refer to figures. Page numbers ending in "t" refer to tables.

A

Absorption, 8, 83, 87
Actinomycetes, 3, 6, 56, 73–79, 81–82
Acute gastrointestinal illness (AGI), 105–106
Adsorption, 56–58
Airborne diseases, 3–4
Airborne pathogens, 6–8, 10, 34–36, 66, 73–79, 82–83
Allergies, 3–4, 80–83
Anaerobic digestion, 113
Animal waste. *See also* Manure
    application of, 113–116
    bacteria in, 67–70, 90–91
    composting, 114–115, 115f
    contamination from, 1–2, 87–93
    diseases from, 87–94
    land application of, 21, 113–116
    parasites in, 90–91
    pathogens in, 6, 44, 87–88, 88f, 88t, 89–94, 92t
    pathways of, 87, 88f
    on plants, 67–70, 68f
    viruses from, 90
Antimicrobial agents, 69, 98
Antimicrobial resistance, 97–98

B

Bacteria
    in air, 10, 36, 76
    in animal waste, 67–70, 90–91
    in soils, 10, 47–50, 50t, 54–57, 55t
Bacterial concentrations, 66–68, 66t, 75–76, 76t
Bacterial infections, 8, 11, 18, 58, 77–79
Bacterial organisms, 34–39, 56–58
Bacterial pathogens, 56–58, 68, 100, 101t, 102t. *See also* Pathogens
Bad Bug Book, 87
Big Necessity, The, 2
Bioaerosols
    actinomycetes, 3, 6, 56, 73–79, 81–82
    endotoxins, 34–37, 73, 79–81
    explanation of, 73–74
    exposure to, 73
    fungi, 33–34, 74t, 75–77, 76t, 78t, 79–83
    glucans, 73, 79, 81
    infections and, 1, 77–79
    inhalation of, 73–75, 74t
    mycotoxins, 73, 77, 79–80, 82–83, 104
    pathogens and, 75–83
    types of, 45, 73–77, 74t, 76t, 78t, 79–83
Biological oxygen demand (BOD), 29
Biological systems, 111
Biosolids. *See also* Sewage sludge
    application of, 10
    composting of, 5
    production of, 41–42, 42t
Botulism, 3, 7, 99
Bowman, Dwight D., 87, 93

C

Centers for Disease Control and Prevention (CDC), 4, 63, 65, 102
Chemical odors, 21–22, 23f, 27–28, 27t
Chemical risk assessment, 16–17. *See also* Risks
Composting procedures, 82, 91, 93, 111–116, 115f
Contamination
    from animal waste, 1–2, 87–93
    of food, 1–2, 44, 67–68, 70
    from human waste, 2–3
    from landfills, 6
    of water, 1–4, 6, 10, 34, 43–44, 48, 56–59, 93–94, 105, 106t

D

Deaths
    from diseases, 2–5, 4t
    from foodborne diseases, 2–3
    from infections, 4
    from waterborne diseases, 3
Decomposition
    decay rates, 51
    of foods, 113
    mycotoxins and, 82–83
    of organic matter, 21, 32–33, 74t, 111–113
Dermal absorption, 8, 10–11, 17–18, 87
Dermal infections, 1, 10–11. *See also* Infections
Diseases. *See also* Infections
    from animal waste, 87–94
    from biosolids, 42t
    deaths from, 2–5, 4t

143

defining, 5
foodborne diseases, 2–4, 47, 67–69, 98–105, 101t, 102t, 103t, 109
gastrointestinal diseases, 31, 39, 98–102, 108
gastrointestinal infections, 2–4, 94, 98–102, 105–106, 108
from human fecal matter, 2–3, 41–46
from landfills, 32–36, 39
malaria, 4, 105, 108
parasitic diseases, 4, 34, 37, 65, 90–91, 101t, 105
from pathogens, 16, 33–37, 35t–36t, 41–45, 42t, 117–141
respiratory disease, 1–4, 39, 73, 99
from sanitation issues, 2–4, 105, 108
from sewage sludge, 2, 41–46, 42t
from solid waste, 1–12, 29–40
from vectors, 37–39, 38t
waterborne diseases, 2–4, 47, 94, 97, 105–109, 106t, 107t, 108t
Disposal
of garbage, 1, 29
of human waste, 41
of municipal solid waste, 1, 29
options for, 111–116
of solid waste, 1, 29, 111–116
Dixie's Forgotten People: The South's Poor Whites, 65
Dose–response
analysis of, 14–15
assessment of, 17f, 18–19, 33
evaluation of, 13
Dumps/landfills
diseases from, 32–36, 39
disposal options for, 111–116
municipal solid waste in, 29
recycling from, 33
runoff from, 6, 29, 32, 34
scavengers at, 9–11, 9f, 13–14, 31–33, 33f, 39
Dust storms, 66, 66t

E

Endotoxins, 34–37, 73, 79–81
Enteric microbial pathogens, 44–45, 45t, 88t
Enteroviruses
microbial pathogens, 44–45, 45t, 51
on plants, 67, 67t
in sludge, 42t
in soils, 50–51, 51t, 57
in water supplies, 105, 106t
Exposure assessment, 13–15, 18–19

F

Flynt, W., 65
Food and Drug Administration (FDA), 14, 87, 98

Food chain, 47–48, 59, 77, 82–83, 93
Food contamination, 1–2, 44, 67–68, 70
Food crops, 47–48, 48f, 59, 67–70, 77, 82–83, 93. *See also* Plants
Food poisoning, 38, 90, 101
Foodborne diseases, 2–4, 47, 67–69, 98–105, 101t, 102t, 103t, 109. *See also* Diseases
Foodborne pathogens, 1–3, 37, 47, 70, 97–98, 102–104, 103t, 109. *See also* Pathogens
Fungal concentrations, 66, 66t, 74t, 75–76, 76t, 79
Fungal infections, 3–4, 11, 77–79, 82. *See also* Infections
Fungi. *See also* Spores
bioaerosols, 33–34, 74t, 75–77, 76t, 78t, 79–83
inhalation of, 8, 66t, 67–69
types of, 45, 76–77, 76t, 78t, 79–83

G

Garbage. *See also* Municipal solid waste
contamination from, 10–11
defining, 5
diseases from, 29–40
disposal of, 1, 29
generation rates of, 29–30, 30f
handling, 29–30, 32–33
health impacts of, 32–33
infections from, 10–11
pathogens in, 29–32
recycling rates of, 30f, 31
Gastrointestinal diseases, 31, 39, 98–102, 108
Gastrointestinal infections, 2–4, 94, 98–102, 105–106, 108
Geophagy
characteristics of, 63–66, 64t
explanation of, 63–65
frequency of, 64–65, 65t
pica and, 8, 63
in plants, 63–71
soils and, 64, 65t
George, Rose, 2
Glucans, 73, 79, 81

H

Hazards
assessment of, 17–18
exposure to, 13
identification of, 13, 17–18
risks and, 13
Hospital wastes, 1, 29. *See also* Municipal solid waste
Household wastes, 29–30, 38–42. *See also* Municipal solid waste
Human enteric microbial pathogens, 44–45, 45t. *See also* Microbial pathogens

Index

Human fecal matter, 2–3, 41–46. *See also* Human waste
Human pathogens. *See also* Pathogens
   bioaerosols and, 75–83
   diseases from, 16, 33–37, 35t–36t, 41–45, 42t
   geophagy and, 63–71
   in plants, 63–71
Human waste
   application of, 113–116
   composting, 114–115, 115f
   contamination from, 2–3
   diseases from, 2–3, 41–46
   disposal of, 41
   land application of, 21, 113–116
   on plants, 67–70, 68f
   water contamination from, 2–3

I

Incineration of wastes, 5, 10, 18, 21, 29–31, 38–41, 112–113, 116
Infections. *See also* Diseases
   bacterial infections, 8, 11, 18, 58, 77–79
   deaths from, 4
   dermal infections, 1, 10–11
   fungal infections, 3–4, 11, 77–79, 82
   gastrointestinal infections, 2–4, 94, 98–102, 105–106, 108
   infective dose data, 33–34, 87–89
   parasitic infections, 4, 65, 90–91, 101t
   respiratory infections, 1–4, 10–11, 73, 81–82, 90
   from solid waste, 1, 8–10, 32–33
   sources of, 8–10
   from vectors, 37–39, 38t
Infective dose data, 33–34, 87–89

L

Land application of wastes, 6, 10, 21, 47–49, 97, 113–116
Landfills
   contamination from, 6
   diseases from, 32–36, 39
   disposal options for, 111–116
   municipal solid waste in, 29
   recycling from, 33
   runoff from, 6, 29, 32, 34
   scavengers at, 9–11, 9f, 13–14, 31–33, 33f, 39
Lead poisoning, 63–64
Lime stabilization, 5–6, 91, 111, 114–116

M

Malaria, 4, 105, 108
Malodors, 21, 24, 26–27. *See also* Odors
Manure. *See also* Animal waste
   application of, 113–116
   bacteria in, 67–70, 90–91
   pathogens in, 6, 44, 87–94
   pathways of, 87, 88f
   on plants, 67, 68f, 69–70
Manure Pathogens, 87, 93
Microbial pathogens, 16, 44–45, 45t, 51
Microorganisms, 54–55, 55t, 56–57
Molds, 3–4, 45, 75–79, 82–83. *See also* Fungi
Municipal solid waste (MSW). *See also* Garbage; Solid waste
   defining, 5
   diseases of, 29–40
   disposal of, 1, 29
   generation rates of, 29–30, 30f
   handling, 29–30, 32–33
   health impacts of, 32–33
   in landfills, 29
   neglect of, 1
   pathogens in, 29–32
   recycling rates of, 30f, 31
Mycotoxins, 73, 77, 79–80, 82–83, 104

N

National Research Council (NRC), 15

O

Odorant, 21–24, 23t, 25f, 27t, 28
Odors
   characterization of, 21–22
   chemical odors, 21–22, 23f, 27–28, 27t
   compounds, 21–22, 25–27, 27t
   concentration of, 21–23, 22f, 23t, 24f
   effects of, 22–28, 24f
   explanation of, 21–22
   health and, 26–27
   hedonic tone of, 22–23, 23f
   intensity of, 21–22, 22f
   malodors, 21, 24, 26–27
   organic odors, 21, 23f, 27t
   persistence of, 22
   quality of, 21–23, 23f
Olfactory system, 21–24, 25f
Organic matter. *See also* Plants; Soils
   decomposition of, 21, 32–33, 74t, 111–113
   odorants and, 21, 27t
   pathogens in, 49, 54–59
Organic odors, 21, 27t
Organisms
   parasites, 4, 51, 52t, 58–59, 65, 90–91
   in soils, 54–55, 55t, 56–59
   survival of, 55t, 56–57

P

Parasites, 4, 51, 52t, 58–59, 65, 90–91, 94
Parasitic diseases, 4, 34, 37, 65, 90–91, 101t, 105

Parasitic infections, 4, 65, 90–91, 101t
Parasitic worms, 51, 52t. *See also* Worms
Pathogens
   absorption of, 8, 83, 87
   airborne pathogens, 6–8, 10, 34–36, 66, 73–79, 82–83
   in animal waste, 6, 44, 87–88, 88f, 88t, 89–94, 92t
   bacterial pathogens, 56–58, 68, 100, 101t, 102t
   bioaerosols and, 75–83
   categories of, 44–45, 45t
   defining, 5
   dermal absorption of, 8, 10–11, 17–18, 87
   diseases from, 16, 33–37, 35t–36t, 41–45, 42t, 117–141
   enteric pathogens, 44–45, 45t, 88t
   in food, 97–110
   foodborne pathogens, 1–3, 37, 47, 70, 97–98, 102–104, 103t, 109
   geophagy and, 63–71
   human enteric microbial pathogens, 44–45, 45t
   in human fecal matter, 2–3, 41–46
   infective dose of, 33–34, 87–89
   ingestion of, 1, 8, 17–19, 31f, 48f, 63–66, 70, 83, 87, 91, 100
   inhalation of, 8, 10, 17–18, 31f, 48f, 73–75, 74t, 80–83, 87
   in manure, 6, 44, 87–94
   microbial pathogens, 16, 44–45, 45t, 51
   microorganisms and, 54–57, 55t
   in organic matter, 49, 54–59
   parasites and, 51, 52t, 58–59
   pathways of, 87, 88f
   on plants, 63–67, 67t, 68–69, 68f, 70–71
   primary pathogens, 44–45
   secondary pathogens, 44–45
   in sewage sludge, 6, 41–46
   in soils, 6–7, 47–62, 50t, 52t, 55t
   in solid waste, 6–11, 7f, 29–40
   sources of, 6–11
   survival of, 9, 47–54, 50t, 51t, 59, 67–68, 67t, 91–94, 92t
   transmission routes of, 47–48, 48f, 87, 88f
   viruses from, 50–53, 51t, 53t, 55t, 57–58
   wastewater treatment process and, 41–43, 43f, 47–49
   in water, 97–110
   waterborne pathogens, 1, 6, 47, 94, 97, 105, 106t, 109
Pica, 8, 63
Plants. *See also* Food crops
   contamination pathways, 67–70, 68f
   enteroviruses on, 67, 67t
   geophagy in, 63–71
   manure on, 67, 68f, 69–70
   pathogens on, 63–67, 67t, 68–69, 68f, 70–71

Q

Quantitative microbiological risk assessment (QMRA), 14

R

Respiratory disease, 1–4, 39, 73, 99
Respiratory infections, 1–4, 10–11, 73, 81–82, 90. *See also* Infections
Risks
   analysis of, 13–15, 15f
   assessment of, 13–14, 15f, 16–19, 17f, 33
   characterization of, 13–15, 15f, 19
   communication of, 13–15
   exposure assessments, 13–15, 18–19
   hazard identification, 13, 17–18
   management of, 13–15, 15f

S

Sanitation
   development of, 1–2
   disease and, 2–4, 105, 108
   importance of, 2
   neglect of, 1
Scavengers, 9–11, 9f, 13–14, 31–33, 33f, 39
Septage, 41–46. *See also* Sewage sludge
Septic system, 42–43
Sewage sludge. *See also* Biosolids
   bacteria in, 42t
   composting of, 5
   diseases from, 2, 41–46, 42t
   land application of, 10, 47–49
   pathogens in, 6, 41–46
   primary sludge, 41
   production of, 29
Sewage treatment system, 41–43, 43f
Soils
   bacteria in, 10, 47–50, 50t, 54–57, 55t
   enteroviruses in, 50–51, 51t, 57
   geophagy and, 64, 65t
   microorganisms in, 54–57, 55t
   organisms in, 54–55, 55t, 56–59
   parasites in, 51, 52t, 58–59, 65
   pathogens in, 6–7, 47–62, 50t, 52t, 55t
   properties of, 6–7, 7f, 47–50
   viruses in, 50–53, 51t, 53t, 55t, 57–58
Solid waste
   anaerobic digestion of, 113
   burning, 112–113, 116
   composting, 114–115, 115f
   defining, 5
   diseases from, 1–12, 29–40
   disposal of, 1, 29, 111–116
   garbage, 29–40
   handling, 29–30, 32–33

health impacts of, 32–33
human fecal matter, 41–46
incineration of, 5, 10, 18, 21, 29–31, 38–41, 112–113, 116
infections from, 1, 8–10, 32–33
land application of, 21, 113–116
lime stabilization of, 5–6, 91, 111, 114–116
management of, 111–116
municipal solid waste, 1, 29–40
pathogens in, 6–11, 7f, 29–40
sewage sludge, 41–46
Spores. *See also* Fungi
bioaerosols, 74t, 75–76, 76t, 77–83
endospores, 98–99
fungal concentrations, 66
inhalation of, 8
Stedman's Medical Dictionary, 5

T

Tertiary treatments, 41, 54
Tetanus, 7

U

United Nations, 31
United States Environmental Protection Agency (USEPA), 13–15, 29, 48

V

Vector-borne diseases, 37–39, 38t
Viruses
adsorption of, 57–58
from animal waste, 90
in soils, 50–53, 51t, 53t, 55t, 57–58
survival of, 53t, 55t

W

Wastewater treatment plants, 21, 29, 41–43, 43f, 47–49, 113
Water
contamination of, 1–4, 6, 10, 34, 43–44, 48, 56–59, 93–94, 105, 106t
enteroviruses in, 105, 106t
parasites in, 94
pathogens in, 97–110
Water treatment facilities, 1, 76
Waterborne diseases, 2–4, 47, 94, 97, 105–109, 106t, 107t, 108t. *See also* Diseases
Waterborne pathogens, 1, 6, 47, 94, 97, 105, 106t, 109. *See also* Pathogens
World Health Organization (WHO), 2, 3, 7, 29, 75, 97, 108
Worms, 4, 6–7, 45t, 51, 52t, 59, 65–66, 105, 108t